aus der Reihe:

Innovationen mit Mikrowellen und Licht

Forschungsberichte aus dem Ferdinand-Braun-Institut, Leibniz-Institut für Höchstfrequenztechnik

Band 51

Mahmoud Tawfieq

Development and characterisation of a diode laser based tunable high-power MOPA system

Herausgeber: Prof. Dr. Günther Tränkle, Prof. Dr.-Ing. Wolfgang Heinrich

Ferdinand-Braun-Institut	Tel.	+49.30.6392-2600
Leibniz-Institut	Fax	+49.30.6392-2602
für Höchstfrequenztechnik (FBH)		
Gustav-Kirchhoff-Straße 4	E-Mail	fbh@fbh-berlin.de
12489 Berlin	Web	www.fbh-berlin.de

Innovations with Microwaves and Light

Research Reports from the Ferdinand-Braun-Institut, Leibniz-Institut für Höchstfrequenztechnik

Preface of the Editors

Research-based ideas, developments, and concepts are the basis of scientific progress and competitiveness, expanding human knowledge and being expressed technologically as inventions. The resulting innovative products and services eventually find their way into public life.

Accordingly, the *"Research Reports from the Ferdinand-Braun-Institut, Leibniz-Institut für Höchstfrequenztechnik"* series compile the institute's latest research and developments. We would like to make our results broadly accessible and to stimulate further discussions, not least to enable as many of our developments as possible to enhance everyday life.

Recent developments in the manufacturing of internal gratings for GaAs-based diode lasers make a wide tuning range up to several ten nanometers possible. This opens new application fields for diode lasers, e.g. as near-infrared light source for sum frequency generation with mid-infrared radiation in up-conversion systems. Such equipment is used to visualize gas clouds, for absorption measurements and wavelength-modulated Raman spectroscopy or even as potential light source for optical coherence tomography. Devices with integrated heater elements showed tuning ranges up to 23.5 nm. When implemented as master oscillator into power amplifier systems, it is possible to generate up to 5 W tunable output power with a narrow spectral linewidth below 17 pm. These features demonstrate the capability of such devices for the targeted applications.

We wish you an informative and inspiring reading

Günther Tränkle

Prof. Dr. Günther Tränkle
Director

Wolfgang Heinrich

Prof. Dr.-Ing. Wolfgang Heinrich
Deputy Director

The Ferdinand-Braun-Institut

The Ferdinand-Braun-Institut researches electronic and optical components, modules and systems based on compound semiconductors. These devices are key enablers that address the needs of today's society in fields like communications, energy, health and mobility. Specifically, FBH develops light sources from the visible to the ultra-violet spectral range: high-power diode lasers with excellent beam quality, UV light sources and hybrid laser systems. Applications range from medical technology, high-precision metrology and sensors to optical communications in space. In the field of microwaves, FBH develops high-efficiency multi-functional power amplifiers and millimeter wave frontends targeting energy-efficient mobile communications as well as car safety systems. In addition, compact atmospheric microwave plasma sources that operate with economic low-voltage drivers are fabricated for use in a variety of applications, such as the treatment of skin diseases.

The FBH is a competence center for III-V compound semiconductors and has a strong international reputation. FBH competence covers the full range of capabilities, from design to fabrication to device characterization.

In close cooperation with industry, its research results lead to cutting-edge products. The institute also successfully turns innovative product ideas into spin-off companies. Thus, working in strategic partnerships with industry, FBH assures Germany's technological excellence in microwave and optoelectronic research.

Development and characterisation of a diode laser based tunable high-power MOPA system

vorgelegt von
M.Sc. Eng.
Mahmoud Tawfieq
geboren in Bagdad

von der Fakultät IV – Elektrotechnik und Informatik
der Technischen Universität Berlin zur Erlangung
des akademischen Grades

Doktor der Naturwissenschaften
- Dr.rer.nat -

genehmigte Dissertation

Vorsitzender: Prof. Dr. Rolf Schuhmann
Gutachter: Prof. Dr. Günther Tränkle
Gutachter: Prof. Dr. Paul Michael Petersen
Gutachter: PD Dr. Bernd Sumpf

Tag der wissenschaftlichen Aussprache: 16. January 2019

Berlin 2019

Bibliografische Information der Deutschen Nationalbibliothek
Die Deutsche Nationalbibliothek verzeichnet diese Publikation in der
Deutschen Nationalbibliografie; detaillierte bibliographische Daten
sind im Internet über http://dnb.d-nb.de abrufbar.
1. Aufl. - Göttingen: Cuvillier, 2019
Zugl.: (TU) Berlin, Univ., Diss., 2019

© CUVILLIER VERLAG, Göttingen 2019
Nonnenstieg 8, 37075 Göttingen
Telefon: 0551-54724-0
Telefax: 0551-54724-21
www.cuvillier.de

ISBN 978-3-7369-9983-1
eISBN 978-3-7369-8983-2

To my beloved parents Alaa Idris Khalil and Tawfieq Al Aybade, without whom I would not be the man I am, not know what I know and not be where I am today.

Preface

This thesis was prepared at the Ferdinand-Braun Institut, Leibniz-Institut für Höchstfrequenztechnik, in fulfilment of the requirements for acquiring a doctorate degree in natural sciences at the Technical University of Berlin.

This work has been carried out under the Mid-TECH project funded by the European Union's Horizon 2020 research and innovation program under Grant Agreement no. 642661.

Mahmoud Tawfieq
Berlin, 16 September 2018

Abstract

The aim of this work is to develop and characterize widely tunable high power diode lasers emitting at 976 nm, to serve in non-linear frequency conversion applications. In particular, to serve as pump sources in infrared upconversion detection systems. This is realized by a master oscillator power amplifier (MOPA) configuration, where the MO provides the wavelength stabilization and tuning, while the PA ensures power amplification to the watt level.

In this thesis, two approaches of developing tunable laser sources are investigated. The first concept utilizes tunable multi-arm distributed Bragg reflector ridge waveguide (DBR-RW) lasers, while the second employ widely tunable sampled-grating (SG) lasers.

In the first approach, three types of monolithic multi-arm lasers are developed: two-arm (Y-branch), four-arm and six-arm DBR-RW lasers. Among the main findings of this work, is that the performance of these lasers is strongly influenced by the intersection point between the individual arms, and is less affected by the bend structure. This is shown in a comparison between three Y-branch lasers with different S-bend (Sinus, Co-sinus and single bend) structures. The performance of these lasers varies only slightly. However, they provide reduced output power and beam quality in comparison to DBR-RW reference lasers. This deterioration is explained by the intersection point between the two laser arms. This is confirmed by passive RW simulations which show that the studied S-bend structures (alone), do not influence the beam quality. Higher order modes are first supported when combining two bend structures, in particular at the intersection point which causes the worse beam quality.

In addition, it is shown that this intersection point also influences the spectral properties of these lasers. When no current is injected into the common section, i.e. becoming absorbent, single mode operation is obtained even at high output powers. This is in contrast to spectral multi-mode operation at high output powers, obtained when this section is injected with current.

Based on the findings of the Y-branch lasers, four-arm lasers with different common sections and arm curvatures are developed and characterized. The first laser have a single intersection point between all four arms, while the second laser have two separate intersection points. The latter structure provides comparable performance from all four arms. The laser with a single intersection point however, shows reduced output power and beam quality for the outer arms in comparison to the inner. This study confirms once again that the laser performance is strongly dependent on the common section and in addition, is less influenced by the arm curvature. This is seen in the comparable performance of the laser with two intersection points, which have stronger curvature for the outer arms in comparison to the inner. This is further approved by results from a six-arm laser that have three individual intersection points. This laser shows less deviation in performance between the six arms, despite the three different curvatures of the outer, middle and the inner arms.

Thermal wavelength tuning of the developed laser sources is obtained by resistor based micro-heaters embedded on top of the grating sections. The characterization of these heaters demonstrate that up to 7.5 nm of tuning can be achieved from each arm of a Y-branch DBR laser. This indicates that a combined tuning of $N \times 7.5$ nm can be obtained from a laser with N arms. This could be achieved by spectrally distancing the gratings of the each arm by 7.5 nm, and is set by the number of arms and ultimately limited by the available gain bandwidth.

The second approach uses SG structures to obtain wide wavelength tuning from a single laser cavity. In this work, numerical tools and design approach are provided to obtain single mode operation over a desired wavelength tuning range. Design parameters such as grating length, period and coupling coefficients are discussed in regards to optimizing the tuning performance. Different vertical structures that can support such gratings are fabricated and characterized, among which a double quantum well

(DQW) based structure, with a 1st order grating and a wide n-cladding layer is proposed. This structure provides coupling coefficients of about 250 cm^{-1}, with a narrow far field angle of 24° at full width at half maximum.

While InP based SG lasers are well established, a milestone of this work is the first time demonstration of fully functional GaAs based SG-lasers. These devices consist of four sections: a gain section, a phase section, a front and a back SG. They emits up to 70 mW of output powers, have a spectral linewidth less than 17 pm with diffraction limited beam quality. They provide up to 21 nm of discrete wavelength tuning when operating a single SG heater, with a SG mode spacing of about 2.3 nm. By operating both SG heaters, 23.5 nm of quasi-continuous tuning is obtained with a mode spacing of about 115 pm.

To increase the output power of the tunable light sources, MOPA systems are developed which utilize tapered power amplifier (TPA) structures. Among the main results of this work, is the development of a hybrid MOPA system, constructed on a compact 25 mm × 25 mm conduction cooled package (CCP). This system provides an output power of 5.5 W from a nearly diffraction limited beam with $M^2_{1/e^2} = 2.2$ along the slow axis. A combined 9.7 nm of wavelength tuning is obtained from the two laser arms. Over this tuning range, a power variation less than 0.5% is observed, while a spectral linewidth small than 17 nm is maintained.

Another main result is the development of a watt level SG-based MOPA system, providing narrow spectral linewidth below 17 pm together with a $M^2_{1/e^2} = 1.6$ along the slow axis. More than 23 nm of quasi-continuous wavelength tuning is achieved, with a power variation between 0.5 and 1.0 W.

The demonstrated tunable MOPA systems provide some of the highest output powers at 976 nm without the use of an external cavity configuration. The capabilities of these devices are demonstrated in some early-stage upconversion experiments. This includes upconversion detection of infrared light between 10 and 12 μm, as well as upconversion based hyperspectral imaging between 6 and 7 μm.

Zusammenfassung

Das Ziel der vorliegenden Dissertation ist die Entwicklung und Charakterisierung von weit abstimmbaren Diodenlaser-basierten Hochleistungslichtquellen bei einer Emissionswellenlänge von 976 nm. Ein mögliches Anwendungsgebiet dieser Laser ist der Einsatz als Pumplichtquelle für die nichtlineare Frequenzkonversion, unter anderem die Upconversion von infrarotem Licht in den sichtbaren Spektralbereich zu ermöglichen.

Grundkonzept dieser Lichtquellen ist eine Master Oscillator Power Amplifier (MOPA) Konfiguration. Der MO realisiert die Wellenlängenstabilisierung und -abstimmung, während der PA die Leistungsverstärkung auf Wattniveau realisiert.

In dieser Arbeit werden zwei Ansätze zur Entwicklung von weit abstimmbaren Laserquellen untersucht. Das erste Konzept verwendet thermisch abstimmbare Bragg-Spiegel Rippenwellenleiter-Laser (DBR-RW-Laser) mit mehreren über Y-Koppler kombinierten Armen. Das zweite Konzept basiert auf einem Sampled-Grating (SG)-Laser, um eine weite Wellenlängenabstimmbarkeit zu erreichen.

Im ersten Ansatz werden drei verschiedene Typen von monolithischen Mehrarmlasern mit einander verglichen: Zweiarmige Y-Laser bilden die einfachste Lösung. Dazu kommen vierarmige und sechsarmige DBR-RW-Laser, die benutzt werden, um den Einfluss unterschiedlicher Krümmungen der Rippenwellenleiter und den Einfluss der Y-Koppler zu untersuchen.

Zu den wichtigsten Ergebnissen dieser Arbeit gehört, dass die Leistung der untersuchten Laser wesentlich von der Form der Kopplungssektion zwischen den einzelnen Armen und weniger von der Biegestruktur beeinflusst wird. Dies wird durch einen experimentellen Vergleich zwischen drei Y-Lasern mit unterschiedlichen S-Biegestrukturen (Sinus, Cosinus und einfache Biegung) gezeigt. Diese Laser erreichen nur geringfügig unterschiedliche Ausgangsleistungen. Allerdings, verschlechtert sich die Strahlqualität und die Ausgangsleistung im Vergleich zu DBR-RW-Referenzlasern deutlich. Diese Verschlechterung kann durch Verluste in der Kopplungssektion zwischen den zwei Laserarmen erklärt werden. Durchgeführte Rechnungen basierend auf einer Simulation für passive Rippenwellenleiter bestätigten dies. Der Einfluss der untersuchten S-Biegestrukturen auf die Strahlqualität ist nur schwach. Räumliche Moden höherer Ordnung, die die Strahlqualität verschlechtern, treten dann auf, wenn zwei Biegestrukturen symmetrisch miteinander kombiniert werden.

Weiterhin wird gezeigt, dass über die Ansteuerung der Kopplungssektion die spektralen Eigenschaften der Laser beeinflusst werden können. Wird die Kopplungssektion nicht elektrisch gepumpt, d.h. sie wirkt absorbierend, bleibt der Grundmode-Betrieb selbst bei hohen Ausgangsleistungen erhalten. Wird diese Sektion elektrisch betrieben, tritt bei hohen Ausgangsleistungen spektraler Multi-Mode-Betrieb auf.

Basierend auf diesen Erkenntnissen wurden vierarmige Laserquellen mit zwei verschiedenen Kopplungssektionen und unterschiedlichen Krümmungsradien prozessiert und ihre elektro-optische, spektralen und Strahleigenschaften miteinander verglichen. Das erste Layout verfügt über einen einzigen Kopplungspunkt an dem sich alle vier Arme treffen, während das andere Design zwei getrennte Kopplungspunkte besitzt. Für letzteren erreichen alle vier Arme eine vergleichbare Leistung, während der Laser mit einem einzigen Kopplungspunkt eine verringerte Ausgangsleistung und Strahlqualität für die äußeren Laserarme im Vergleich zu den inneren Armen zeigt. Diese Untersuchung zeigt erneut, dass die Laserleistung primär von der Art der Kopplung abhängt und weniger von der Biegestruktur. Dies bestätigen Messungen für die Lichtquellen mit zwei getrennten Kopplungspunkten. Trotz unterschiedlicher Krümmungsradien wird für alle Arme die gleiche Ausgangsleistung erreicht. Weitere Untersuchungen an den sechsarmigen Lasern mit getrennten Kopplungspunkten zeigten dieses Ergebnis. Trotz der drei unterschiedlichen Krümmungsradien der äußeren, mittleren und inneren Arme, sind nur geringere Unterschiede in der Laserleistung der sechs Arme festzustellen.

Die spektrale Abstimmung der Laser wird über resistive Mikroheizer realisiert, die oberhalb der DBR-Gitter in die Gittersektionen eingebettet sind. Durch Betreiben der Heizelemente ist eine Durchstimmung von bis zu $7,5$ nm pro Arm erreicht worden. Für einen Laser mit N Armen ergibt sich damit, wenn der Abstand der prozessierten Gitterwellenlängen $7,5$ nm beträgt, eine Abstimmung von $N \times 7,5$ nm. Die Gesamtabstimmbarkeit ist dann durch die Anzahl der Arme bestimmt und durch die Verstärkungsbandbreite limitiert.

Beim zweiten Ansatz wird ein SG-Laser untersucht, der eine große Wellenlängenabstimmung eines einzelnen Lasers ermöglicht. Voraussetzung für das Design dieser Laser mit dem gewünschten abstimmbaren Wellenlängenbereich waren umfangreiche numerische Simulationen. Die verschiedenen Designparameter (Gitterlänge, Periode und Kopplungskoeffizient) werden im Hinblick auf die Optimierung der Laserleistung und der Durchstimmung diskutiert. Geeignete vertikale Schichtstrukturen, die die erforderlichen Gitterstrukturen ermöglichen, wurden hergestellt und charakterisiert. Favorisiert wurde dabei eine Schichtstruktur mit doppelten Quantengraben sowie einem Gitter erster Ordnung und einer breiten n-Mantelschicht. Diese Struktur liefert die notwendigen Kopplungskoeffizient von ungefähr 250 cm^{-1} zusammen mit einem geringen vertikalen Fernfeldwinkel von 24° (Halbwertsbreite).

SG-Laser wurden bisher nur auf InP Basis realisiert. Erstmalig konnte im Rahmen dieser Arbeit funktionsfähige GaAs-basierte SG-Laserquellen demonstriert werden. Die entwickelten Laser bestehen aus vier Abschnitten: einer Gewinn-, einer Phasen-, einer vorderen und einer hinteren SG Sektion. Diese liefern Ausgangsleistungen von bis zu 70 mW und haben eine spektrale Linienbreite kleiner als 17 pm mit nahezu beugungsbegrenzter Strahlqualität. Auch hier erfolgt die Abstimmung mittels oberhalb der Gitter implementierter Heizer. Es konnte bei Betrieb eines einzelnen SG-Heizelementes eine diskrete Wellenlängenabstimmung von bis zu 21 nm erzielt werden. Der SG-Modenabstand ist hierbei etwa $2,3$ nm. Bei Verwendung beider SG Heizelemente wird eine quasi-kontinuierliche Abstimmung über $23,5$ nm möglich, mit einem Modenabstand von 115 pm.

Um die Ausgangsleistung der abstimmbaren Lichtquellen zu erhöhen, wurden MOPA-Systeme entwickelt, bei denen Trapez-Verstärker verwendet wurden. Eines der Hauptergebnisse dieser Arbeit ist ein hybrides Y-Laser-basiertes MOPA-System auf einer Grundfläche von nur 25 mm × 25 mm auf einer Conduction Cooled Package (CCP) aufgebaut. Das MOPA-System liefert eine Ausgangsleistung von $5,5$ W mit einer nahezu beugungsbegrenzten Strahlqualität mit einer Beugungsmaßzahl $M_{1/e^2}^2 = 2,2$ in der lateralen Achse. Der MOPA kann entsprechend der spektralen Eigenschaften des MO über $9,7$ nm abgestimmt werden. Über den gesamten Abstimmbereich ist die Leistungsvariation kleiner als $0,5\%$, mit einer spektralen Breite von kleiner als 17 pm.

Ein weiteres MOPA-System wurde auf Basis von SG-Lasern entwickelt. Mit diesen können Leistungen im Wattniveau bei einer Strahlqualität von $M_{1/e^2}^2 = 1,6$ in der lateralen Achse erreicht werden. Wie auch der verwendete MO, lässt sich der MOPA spektral über 23 nm abstimmen bei einer spektralen Breite von unter 17 pm. Hier variiert die Ausgangsleistung über den Abstimmbereich zwischen $0,5$ W und $1,0$ W.

Die entwickelten abstimmbaren MOPA-Systeme erreichen die höchsten publizierten Ausgangsleistungen bei 976 nm ohne die Verwendung einer externen Kavität. Das Potenzial dieser Systeme wird in Upconversion-Experimenten zur Detektion von Infrarot Strahlung zwischen 10 bis 12 μm, sowie Upconversion-basierter hyperspektraler Bildgebung zwischen 6 und 7 μm, demonstriert.

Acknowledgement

First and above all, I praise and thank **God**, the almighty for giving me the strength, knowledge, ability and opportunity to undertake this research study and to persevere and complete it satisfactorily.

Next, I would like to take a moment and express my gratitude and acknowledge the many people who made this thesis possible.

The main contributor is my dear supervisor PD. Dr. Bernd Sumpf whom I'm forever indebted and grateful for. Bernd guided and supervised me through this PhD project and shares the success of this project. Beside the academic supervision, Bernd also taught me about international co-operation, scientific planning and strategy. His many strict comments and corrections ensured that my papers, slides and presentations were always of high standard. Secondly, I would like to thank Dr. Hans Wenzel for his guidance and supervision of the numerical simulations performed during this project.

Of course I must thank the man who made this possible, my main supervisor Prof. Günther Tränkle who offered me this position and supported me through the ups and downs of this project. From the first day, I have been inspired by Günthers high working moral, ambition and unique leadership. Therefore I hope to take, or at least develop some these characteristics along my personal and professional journey.

Next, a handful of people deserves similar acknowledgement for their outstanding technical contributions which made this project possible: Jörg Fricke for the manufacturing of the multi-arm lasers, Pietro Della Casa for growing the different implemented wafers, Olaf Brox for the manufacturing of the sampled-gratings, Peter Ressel for the facet coating, David Feise for providing the tapered power amplifiers, Arnim Ginolas, Sabrina Kreutzmann, Felix Müller and Marvin Schilling who all helped with the mounting of the lasers. Without their professional assistance and effort, this thesis would not have been possible in any way. No lasers = no thesis = no fun.

I would also want to thank my current and previous colleagues at the Laser Sensors Lab; André Müller, Christof Zink, Julia Mewes, Lara Sophie Theurer, Marcel Braune, Martin Maiwald and Norman Ruhnke, for helping me settle down in Berlin and at FBH, and for providing a pleasant working environment.

A special thanks goes to Prof. Markus Weyers for providing excellent proofreading, corrections and comments to all of my publications. Likewise, I would like to thank Kerstin Arzberger-Sumpf for proofreading the Zusammenfassung section of this work.

In general, I would thank all my colleagues at FBH for providing a good and friendly working environment. In particular, I would like to thank Manuela Münzelfeld for helping me with travel bookings and with different bureaucratic stuff. I would also like to thank my colleagues and friends Maruf Hossain, Tanjil Shivan, Basem Arar, Mohamed Brahem and Mohamed Elattar who made my stay at FBH very pleasant.

I would also like to thank my colleagues from DTU. Mainly my co-supervisors Christian Pedersen and Peter Tidemand-Lichtenberg, who not only supervised me in this project but also were my Lecturers in various of my Bachelor and Master courses, and in fact shaped my way of understanding optics and photonics. Also from DTU, I would like to thank Saher Junaid, Ajanta Barh and Yu-Pei Tseng for the upconversion experiments performed at DTU Fotonik in Roskilde.

Similar gratitude deserve the colleagues at ICFO in Barcelona, in particular Prof. Majid Ebrahim-Zadeh, Chaitanya Kumar Suddapalli and Anuja Padhye for our co-operation during my external stay, where the initial tests of realizing a diode pumped OPO system were done.

Finally, thanking my parents and siblings for their support is also a must, although words can not really credit their support. They have shared every pleasant, sad and difficult moment of this project, and have always been supportive and encouraging.

List of publications

Peer-reviewed journal publications

1. **M. Tawfieq**, H. Wenzel, O. Brox, P. Della Casa, B. Sumpf and G. Tränkle, "Concept and numerical simulations of a widely tunable GaAs-based sampled grating diode laser emitting at 976 nm", *IET Optoelectronics* 11 (2), pp. 73-78, January 2017.

2. O. Brox, **M. Tawfieq**, P. Della Casa, P. Ressel, B. Sumpf, G. Erbert, A. Knigge and G. Tränkle, "Realisation of a widely tuneable sampled grating DBR laser emitting around 970 nm", *Electronics Letters* 53 (11), pp. 744-746, May 2017.

3. **M. Tawfieq**, A. Müller, J. Fricke, P. Della Casa, P. Ressel, A. Ginolas, D. Feise, B. Sumpf and G. Tränkle, "Compact High Power Diode Laser MOPA System with 5.5 nm Wavelength Tunability", *IEEE Photonics Technology Letters* 29 (22), pp. 1983-1986, November 2017.

4. **M. Tawfieq**, A. Müller, J. Fricke, P. Della Casa, P. Ressel, D. Feise, B. Sumpf and G. Tränkle, "Extended 9.7 nm tuning range in a MOPA system with a tunable dual grating Y-branch laser", *Optics Letters* 42 (20), pp. 4227-4230, October 2017.

5. **M. Tawfieq**, J. Fricke, A. Müller, P. Della Casa, P. Ressel, A. Ginolas, H. Wenzel, B. Sumpf and G. Tränkle, "Characterization and comparison between different S-bend shaped GaAs Y-branch DBR lasers emitting at 976 nm", *Semiconductor science and technology* 33 (11), p. 115001, October 2018.

6. **M. Tawfieq**, H. Wenzel, O. Brox, P. Della Casa, A. Knigge, M. Weyers, B. Sumpf and G. Tränkle, "High power sampled-grating based MOPA system with 23.5 nm wavelength tuning around 970 nm", *Applied optics* 57 (29), pp. 8680-8685, October 2018.

7. **M. Tawfieq**, J. Kabitzke, J. Fricke, P. Della Casa, P. Ressel, A. Ginolas, H. Wenzel, B. Sumpf and G. Tränkle, "Characterization and comparison between two coupling concepts of four-wavelength monolithic DBR ridge waveguide diode laser at 970 nm", submitted to *Applied physics B*, August 2018.

8. **M. Tawfieq**, J. Fricke, C. Stölmacker, P. Della Casa, B. Sumpf and G. Tränkle, "Spatial filtering of a 6-wavelength DBR-RW laser in a MOPA system", submitted to *IEEE Journal of Selected Topics in Quantum Electronics*, January 2019.

Conference proceedings

1. **M. Tawfieq**, H. Wenzel, B. Sumpf and G. Tränkle, "Concept and numerical simulations of widely tunable GaAs-based sampled-grating diode laser emitting at 976 nm", *Semiconductor and Integrated Optoelectronics Conference: SIOE'16*, Cardiff University, Cardiff, Wales, April 2016.

2. P. Della Casa, O. Brox, A. Knigge, B. Sumpf, **M. Tawfieq**, H. Wenzel, M. Weyers and G. Tränkle, "Integration of active, passive and buried-grating sections for a GaAs-based, widely tunable laser with sampled grating Bragg reflectors", Compound Semiconductor Week 2017 (CSW), Berlin, Germany, May 2017.

3. **M. Tawfieq**, H. Wenzel, O. Brox, P. Della Casa, A. Knigge, M. Weyers, B. Sumpf and G. Tränkle, "Design and realization of a widely tunable sampled-grating distributed-Bragg reflector (SG DBR) laser emitting at 976 nm", *CLEO Europe 2017*, Munich, Germany, July 2017.

4. **M. Tawfieq**, A. Müller, J. Fricke, P. Della Casa, P. Ressel, A. Ginolas, D. Feise, B. Sumpf and G. Tränkle, "972 nm Diode Laser MOPA System with 6.2 W Output Power and a Tuning Range of 5.5 nm", *The 24th Congress of the International Commission for Optics (ICO-24)*, Tokyo, Japan, August 2017.

5. **M. Tawfieq**, H. Wenzel, O. Brox, P. Della Casa, A. Knigge, M. Weyers, B. Sumpf and G. Tränkle, "High power widely tunable sampled-grating based MOPA laser system", DOPS Annual Meeting, Kongens Lyngby, Denmark, November 2017.

6. O. Brox, P. Della Casa, **M. Tawfieq**, B. Sumpf, A. Knigge, M. Weyers and H. Wenzel, "Reflectors and tuning elements for widely-tuneable GaAs based sampled grating DBR lasers", Proc. SPIE 10553, Novel In-Plane Semiconductor Lasers XVII, 1055310, San Francisco, USA, February 2018.

7. **M. Tawfieq**, A. Müller, J. Fricke, P. Della Casa, P. Ressel, A. Ginolas, D. Feise, B. Sumpf and G. Tränkle, "5.5 nm wavelength tunable high power MOPA diode laser system at 971 nm", Proc. SPIE 10553, Novel In-Plane Semiconductor Lasers XVII, 105531F, San Francisco, USA, February 2018.

8. **M. Tawfieq**, A. Müller, J. Fricke, P. Della Casa, P. Ressel, A. Ginolas, D. Feise, B. Sumpf and G. Tränkle, "High power Y-branch MOPA-system with 9.7 nm wavelength tunability for IR up-conversion detection", *High-brightness Sources and Light-driven Interactions Congress - Mid-Infrared Coherent Sources*, Strasbourg, France, March 2018.

9. **M. Tawfieq**, H. Wenzel, O. Brox, P. Della Casa, A. Knigge, M. Weyers, B. Sumpf and G. Tränkle, "Widely tunable high power sampled-grating MOPA system emitting around 970 nm", *CLEO Pacific Rim 2018 Conference*, Hong Kong, China, July 2018.

10. A. Barh, **M. Tawfieq**, B. Sumpf, C. Pedersen and P. Tidemand-Lichtenberg, "Upconversion on Demand using a 2-D Quasi Phase Matching Crystal", *International Conference on Fiber Optics and Photonics - PHOTONICS 2018*, New Delhi, India, December 2018.

Other contributions and highlights

1. Student travel award, at The International Society for Clinical Spectroscopy (CLIRSPEC) Summer School, Lake Windermere, United Kingdom, July 2017.

2. OSA/SPIE Student Paper Award, at 24th congress of the International Commission for Optics (ICO-24), Tokyo, Japan, August 2017.

3. M. Tawfieq, "Diode lasers & systems for spectroscopic applications", oral presentation at the *OpTecBB Summer School 2017 - photonic Integration within academia and industry*, Berlin, Germany, September 2017.

4. M. Tawfieq, "Breaking the wall of applied infrared spectroscopy", 3 minutes long oral presentation at the *Falling Walls Lab science competition in Berlin, Adlershof*, which was awarded the second place at the competition, October 2017.

5. Best poster award at *DOPS annual meeting 2017, Danish optical society*, Kongens lyngby, Denmark, November 2017.

6. Two-page interview by Catarina Pietschmann "Durch Addition zu mehr Empfindlichkeit / Enhancing sensitivity by addition", Leibniz Forschungsverbund Berlin e.v., Verbundjournal, pp. 16-17, December 2017.

List of acronyms

AR Anti-reflection

ASE Amplified spontaneous emission

ASLOC Asymetric super-large optical-cavity

BA Broad area

BM Back mirror

CCP Conduction cooled package

COMD Catastrophic optical mirror damage

CW Constant wave

DBR Distributed Bragg reflector

DFB Distributed feedback

DQW Double quantum well

DWDM Dense wavelength division multiplexed

EDFA Erbium-doped fiber amplifier

FAC Fast axis collimation

FM Front mirror

FSR Free spectral range

FTIR Fourier-transform infrared spectroscopy

FWHM Full width at half maximum

IR Infrared

LED Light emitting diode

LIDAR Light detection and ranging

LOC Large optical cavity

MEMS Microelectromechanical systems

MOPA Master oscillator power amplifier

MOVPE Metal organic vapour phase epitaxy

MQW Multi quantum well

NA Numerical aperture

OCT Optical coherence tomography

OPO Optical parametric oscillator

PUI Power-voltage-current

QCL Quantum cascade laser

QW Quantum well

RMS Repeat mode spacing

RW Ridge waveguide

SAC Slow axis collimation

SB Single bend

SDOM Standard deviation of mean

SEM Scanning electron microscope

SERDS Shifted excitation Raman spectroscopy

SFG Sum frequency generation

SG Sampled-grating

SHG Second harmonic generation

SIMS Secondary ion mass spectrometry

SMSR Side mode suppression ratio

SQW Single quantum well

TE Transverse electrical

TM Transverse magnetic

TPA Tapered power amplifier

VCSEL Vertical-cavity surface-emitting laser

WMRS Wavelength modulated Raman spectroscopy

Chemical compounds

Al aluminium

AlGaAs aluminium gallium arsenide

AlN aluminium nitride

AuSn gold tin

CBr$_4$ tetrabromomethane

CuW copper tungsten

GaAs gallium arsenide

GaAsP gallium arsenide phosphide

InGaP indium gallium phosphide

InGaAs indium gallium arsenide

InP indium phosphide

Nd:YVO$_4$ neodymium-doped yttrium orthovanadate

O oxygen

PbSn lead tin

ZnSe zinc selenide

Contents

Chapter 1

Introduction

In May 1962, Theodore Harold Maiman demonstrated the first laser operation from a solid state ruby laser [1]. In the following months, extensive research took place on stimulated emission and already by November 1962, several groups began reporting on lasing action in semiconductors [2, 3]. Semiconductor based diode lasers have since been improved in terms of manufacturing, output power, beam quality and spectral coherence [4]. They offer lasing emission ranging between the visible and into the long infrared (IR) wavelengths [5]. Due to their compact size, reliability and cheap manufacturing costs, diode lasers have replaced many solid state, chemical and gas lasers.

These advantages have made diode lasers by far the most common laser type, with several million devices being produced every month. Today, semiconductor diode lasers are found in CD players, laser writers, laser pointers and in the bar code readers in the supermarket. In addition, various medical and industrial fields utilizes diode lasers, and still more and more applications are found [6].

1.1 Tunable diode lasers

The emission wavelength of most, if not all laser sources can in principle be changed by different means. In this work however, tunable refer to laser sources with a mechanism for precise control of the emission wavelength while maintaining a narrow spectral linewidth. Tunable diode lasers holds a strong foothold within various applications. Among the main applications is optical telecommunication, where wavelength tuning is one of many methods to increase the bandwidth transmitted through optical fibers [7, 8, 9].

In addition, tunable lasers especially emitting in the near IR are key components within the field of bio-medical optics, including applications such as Fourier domain optical coherence tomography (OCT) [10, 11], absorption spectroscopy [12, 13] and wavelength modulated Raman spectroscopy (WMRS) [14, 15]. Likewise, applications such as trace-gas detection [16], non-linear frequency conversion [17, 18] and terahertz frequency generation [19, 20] all utilize tunable laser sources.

The specific light source requirements and tuning capabilities varies depending on the targeted application. In some applications, say in OCT, wide wavelength tuning at rapid speeds is needed for obtaining high axial resolution scans at high imaging rates. On the other hand, only tens of milliwatts are needed as you do not want to harm the patients retina during imaging.

Other applications require moderate wavelength tuning but with high output powers. This is e.g. the case in differential absorption light detection and ranging (LIDAR) applications [21], where high output powers are needed to get sufficient back scattered photons at the detector. On the other hand, wavelength tuning should only cover on/off resonance absorption peaks and has instead specific linewidth requirements set by the absorption peaks [22].

1.2 Upconversion IR detection

Spectral detection of IR wavelengths is of great technical and scientific interest, due to the important chemical compounds that display unique and strong IR spectral fingerprints [23]. In addition, IR imaging is a valuable tool to investigate biological samples, such as histological tissue samples and cell cultures [24, 25]. The non-invasive detection can provide chemical specificity without recourse to labels, and is currently used to understand and identify different types of cancer [26].

In this work, the developed light sources are targeted applications within the field of non-linear frequency conversion, in particular upconversion IR detection. Recently, upconversion detection has had a renaissance due to the progress in the field of IR light sources and non-linear materials. The upconversion technique shifts IR signals into the near IR range, where efficient detection can be made with basic Si-based detectors at room temperature [23]. In comparison, direct detection using traditional IR detectors, e.g. HgCdTe detectors, require cryogenic cooling in order to improve the noise performance [27, 28]. Nonetheless, noise remains a major issue in these detectors, as all objects at room temperature have their Black-body radiation in the IR range, including the detector itself.

In the case of upconversion detection, this problem is solved by first converting the IR signals toward shorter wavelengths, before being detected by efficient and low noise detectors. This technique has proven superior noise performance [23, 29], including single-photon detection at room temperature [30, 31]. In addition, IR hyperspectral imagining has been demonstrated with this technique [32, 33] as well as LIADR applications [34, 35].

Upconversion is based on the non-linear process of sum frequency generation (SFG), where the IR signal is mixed with a pump source inside a non-linear crystal. Under phase matching condition, this result in generation of an upconverted signal with the frequency $\nu_{\text{SFG}} = \nu_{\text{signal}} + \nu_{\text{pump}}$ [36], i.e. a signal with a shorter wavelength of

$$\lambda_{\text{SFG}} = \left(\frac{1}{\lambda_{signal}} + \frac{1}{\lambda_{pump}} \right)^{-1} = \frac{\lambda_{\text{signal}} \lambda_{\text{pump}}}{\lambda_{\text{signal}} + \lambda_{\text{pump}}} \ . \tag{1.1}$$

In addition, to upconvert a wide IR spectrum, the angle of incident at the non-linear crystal or the temperature of the crystal can be changed. By doing so, different phase matching conditions are fulfilled providing upconversion of a broad IR spectrum [37]. Alternatively, and as is intended by this work, by tuning the wavelength of the pump source, different phase matching conditions are fulfilled, providing broad IR upconversion enabling spectroscopy and hyperspectral imaging in the IR [32, 33, 38, 39]. This could provide a quicker method of IR sweeping/detection, where the width of the upconverted IR spectrum will be set by the bandwidth of the upconversion process, together with the total tunability of the pump source.

In summary, the wider the pump wavelength tuning the broader an IR signal can be upconverted, thus addressing more IR features. Moreover, as this is a non-linear process and as the IR signals are expected to be of low power, high pump powers are needed to improve the upconversion efficiency.

1.3 Types of tunable diode lasers

Semiconductor based distributed Bragg reflector (DBR) and distributed feedback (DFB) ridge waveguide (RW) lasers can provide diffraction limited beams, moderate wavelength tuning (< 10 nm) and emit output powers below the watt level [40, 41]. Recently, single wavelength broad-ridge waveguide DBR lasers has been demonstrated with a couple of watts of output power [42].

On the contrary, tapered diode lasers with integrated DBR gratings provide multiple watts of diffraction limited output powers [43], however their tuning capabilities have not been demonstrated. This is mainly due to the complicated structure of these devices, including the challenges of astigmatisms and of obtaining single mode operation. One approach combines tapered power amplifiers (TPA)s in external cavity configurations, which can provide both tuning and high output powers [44, 45]. Such a configuration utilizes a grating which is mechanically adjusted to select and tune the emission wavelength.

Recently, demonstration of electrically pumped tunable vertical cavity surface emitting (VCSEL) lasers with a micro-electro-mechanical system (MEMS) has proved more than 50 nm of tuning at 400 kHz sweeping rates, which makes them ideal for OCT applications [46, 47]. The drawback of this type of laser is the low (< 10 mW) output powers which limits their applications.

The described tunable lasers offers some challenges such as the limited tuning range of DBR and especially DFB lasers, and the low output powers of the VCSEL lasers. In addition, many applications requires continuous tuning which is not typically achieved from the GaAs based DBR and DFB lasers.

1.4 Motivation

The motivation of this work is to develop tunable high power laser sources to serve as compact pump sources in upconversion detection systems. These should act as single-pass pump sources, and replace the standard Nd:YVO$_4$ laser cavities oscillating at 1064 nm that are typically implemented in upconversion systems [23, 35].

The wavelength of the developed diode lasers will be around λ_{pump} = 976 nm, chosen so that the upconverted signals will be near the absorption peak of Si-detectors, i.e. $800 - 900$ nm [48]. Although the difference between 1064 nm and 976 nm is not that significant, the shorter pump wavelength will have a major impact, especially when upconverting longer IR signals ($> 10 \ \mu$m). E.g. assuming a signal around 10 μm [49, 50], the chosen pump wavelength will upconvert the signal to a wavelength of λ_{SFG} = 890 nm, see eq. (1.1), which is ideal for Si-based detectors. In addition, replacing a cavity with a single-pass system will reduce the complexity and footprint of the upconversion detection modules, and potentially allow hand-held IR detectors.

In addition, having a wavelength tunability of about 23 nm (major result of this work), an IR sweep of about 243 cm^{-1} can be obtained which can cover whole spectrum of rotational vibrational bands [51]. In terms of hyperspectral imaging, the high output power and the specific wavelength ensures efficient detection (with reduced integration time), and the wide wavelength tunability will provide an increased field of view [33].

While continuous wavelength tuning is important in many spectroscopic applications, mode hops can be tolerated in upconversion detection as long as they are below the nanometer range, which is the typical conversion bandwidth. The investigated light sources of this work provide moderate (< 0.5 W) output powers, while the targeted application requires high pump powers. Therefore, these light sources are implemented in miniaturized master oscillator power amplifier (MOPA) configurations, where the MO provides the wavelength tunability and the PA increases the output power.

The target of this work, beside the development and understanding tunable laser devices, is to reach suitable output powers for upconversion application with wide wavelength tuning around 976 nm.

1.5 Project Scope

This thesis describes shortly some of the theory behind tunable diode lasers in chapter 2, followed by a description of the first laser process, including two reference DBR-RW lasers in chapter 3. This work examine two concepts of achieving wide wavelength tunability. The first consider using monolithically combined tunable multi-wavelength DBR-RW lasers to cover a wider tuning range compared to a single tunable DBR-RW laser, see chapter 4 and 5. In this case, thermal wavelength tuning is obtained by embedding micro-heaters on top of the grating sections as described chapter 6. The second concept consider using tunable sampled-grating (SG) DBR-RW lasers, which achieve wide wavelength tunability due to their unique grating features, see chapter 7.

Both investigations include the whole "chain" of development, including laser design, device processing, characterization and optimization, all done with regards to the intended MOPA implementation and targeted application. This MOPA implementation is described in chapter 8 where three different MOPA systems are constructed. These systems have different tunable MO lasers, and utilize TPAs to obtain high output powers.

Chapter 9 describes some early-stage upconversion experiments where the developed MOPA systems have been implemented and tested.

Finally, an outlook is given in chapter 10 describing the further development of this work, in particular different strategies to deal with the limits and challenges of achieving wider wavelength tuning, and higher output powers than those obtained in this work.

Chapter 2

Fundamentals of laser diodes

In this chapter, fundamental concepts of semiconductor lasers will be briefly introduced, as required for understanding laser operation, laser gain and wavelength tuning. The reader is here assumed to have a basic understanding of semiconductor physics. These concepts will be used to describe the relevant device characteristics at later chapters. This chapter begins with the concept of optical gain, with focus on semiconductor heterostructures. This is followed by laser threshold, lateral waveguiding and mode spectrum. The presented theory and the figures of this chapter are inspired by the text books [5, 41, 52].

2.1 Optical Gain in Semiconductors

Semiconductor diode lasers are based on energy bands, with the lower band referred to as the valence band and the upper band referred to as the conduction band. These bands are separated by a gap with energy E_g, and for typical III-V compound semiconductors, this gap is in the range of 0.5 to 2.5 eV, depending on the material composition. The photon energy is defined as [5]

$$E_{\text{photon}} = h\nu = \frac{hc}{\lambda} \, , \tag{2.1}$$

where $h = 4.135 \times 10^{-15}$ eV · s is Plancks constant, ν is the frequency of the light and $c = 3 \times 10^8$ m/s is the speed of light in vacuum. The corresponding wavelengths of the aforementioned band gap energies lie in the range between the blue to the near IR [5].

In the absence of an external pump source and at a temperature $T = 0$ K, the valence band of undoped semiconductors is occupied with electrons while the conduction band will be empty. This is reversed by applying an injection current I_{inj} to the semiconductor material, causing an increase in the concentration of electrons in the conduction band and holes in the valence band. This provides generation and recombination between the electron-hole pairs, leading to four basic electronic transitions. These are illustrated in Fig. 2.1 and include:

(a) Spontaneous recombination (photon emission)

(b) Stimulated generation (photon absorption)

(c) Stimulated recombination (coherent photon emission)

(d) Non-radiative recombinations

The open circles in Fig. 2.1 represent unfilled states (holes) and the solid circles represent filled states (electrons).

First, spontaneous recombination of electrons and holes results in the emission of photons of a wavelength equal to the energy difference between the corresponding levels in the conduction and valence bands. This being said, the emitted photons are random in phase and direction. This emission occurs at a spontaneous transition rate of R_{sp} and is utilized in light-emitting diodes (LED)s.

5

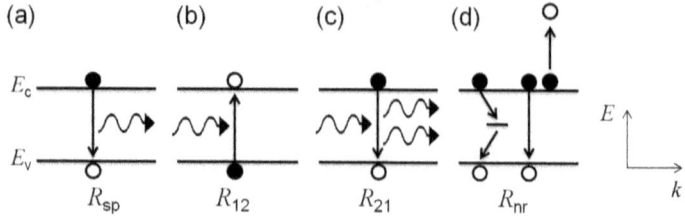

Figure 2.1: *Electronic transitions between the conduction and valence bands showing (a) spontaneous emission (b) photon absorption from level 1 to 2 (c) stimulated emission from level 2 to 1 and (d) two non-radiative recombination processes.*

Photons can also be absorbed by the material which causes generation of new pairs of electrons and holes with a rate of R_{12}, see Fig. 2.1(b). The third transition is stimulated emission, where the generated photons stimulate recombination of additional electrons and holes with a rate of R_{21}. This generation causes simultaneous generation of additional photons that are coherent with the initial photons, in terms of phase and direction, see fig. 2.1(c). This transition is mandatory for laser operation and sets up the two requirements for lasing.

The first requirement is population inversion, i.e. having the number of electrons in the conduction band exceeding the number of holes in the valence band. This is obtained by means of pumping, which in diode lasers is typically done by an electron current. This condition will be discussed further in Sec. 2.4. The second requirement is feedback, which is obtained by positioning the laser medium inside a resonant cavity. By doing so, a number of photons circulate the cavity causing coherent photon emission. In the case of diode lasers, the cavity is typically constructed between coated end facets, with back facets having high reflectivity ($R_b > 90\%$) and the front facet typically having anti-reflection (AR) coating to have high output power. This causes most of the light to be emitted from the front facet where it is typically collected and collimated. Alternatively, and as will be utilized in the next chapters, the back mirror can be made using DBR gratings, which provide high reflectivity in a narrow spectral range leading to a narrow emission linewidth.

The final electron transition is non-radiative, with a rate R_{nr} at which electron-hole pairs are recombined without emission of any photons, through e.g. traps and/or Auger recombination [52], see fig. 2.1(d). This process is unwanted in diode lasers as it reduces the efficiency of the light source and therefore, different design concepts are implemented to reduce these effect.

2.2   Vertical confinement

To fulfill the two laser requirements described in the previous section, diode lasers require a medium providing optical gain, an optical waveguide structure confining photons to the active region and a lateral confinement of photons providing injection current and carriers [5].

Most diode lasers are based on p-i-n double heterostructures consisting of an undoped semiconductor layer embedded between p- and n-doped semiconductors with larger band gaps. More specifically, in this work a graded index separate confinement heterostructure (GRINCH) [53] is used, see Fig. 2.2(a). This structure (GaAs based) has the lowest Al doping level near the active region ($\sim 10^{16}$ cm^{-3}) to reduce absorption effects, while the outer contact and substrate layers have higher doping levels between 10^{18} and 10^{19} cm^{-3} to allow efficient carrier generation and low resistance [54]. A graded doping transition is made between the outer contact and substrate layers and into the active region.

Such structures have the advantage of a narrow carrier confinement (active) region, providing a high recombination rate, separated from the wider graded optical waveguide region, see Fig. 2.2(b). The active region can be realized using single- or multi-quantum well (SQW, MQW). This structure allows optimization of the optical confinement without affecting the carrier confinement. Particularly, smaller band gap semiconductors have higher refractive indices, where the optical field will be confined and

6

guided, see fig. 2.2(c) and (d). This improved confinement and guiding of a GRINSCH structure provide reduced threshold current and higher output powers [53, 55].

The different vertical structures used in this thesis including the active materials will be described in the next chapters.

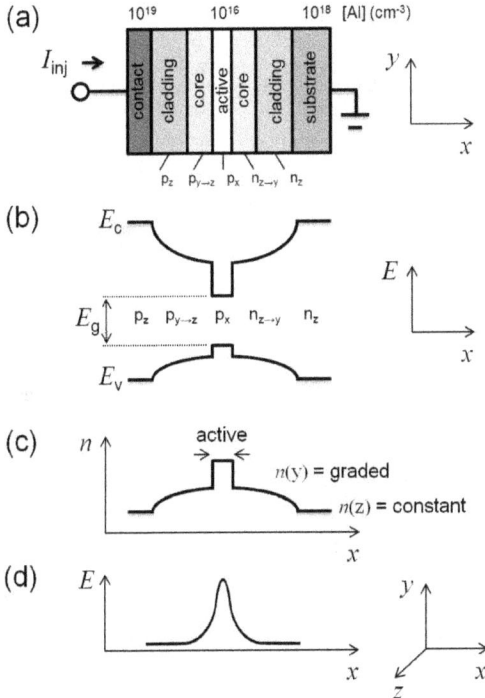

Figure 2.2: *Schematic of the GRINSCH structure showing (a) material structure of the individual layers (b) energy diagram of the conduction E_c and valence E_v band (c) refractive index profile and (d) corresponding electric field profile for a mode travelling in the z-direction.*

2.3 Lateral confinement

Semiconductor diode lasers utilize different concepts for lateral confinement. One concept is gain guiding, which is enabled by defining a current aperture in the contact layer that spatially limits the carrier injection [5]. This means that optical waves outside this aperture will experience high losses, resulting in a lateral confinement of the generated light. This concept is typically used in broad area (BA) lasers where stripe widths of a few hundred microns enable tens of watts of output power [56]. The main disadvantage of BA lasers is the poor beam quality, caused by the high-order lateral modes that can fit within the wide active region [5].

Improved beam quality can be obtained when using narrow stripe width together with the concept of index guiding, which is based on an induced refractive index step in the lateral direction. Similar to an optical fiber, an induced refractive index step provides guiding and confinement in the lateral direction [5].

7

Assuming a three-layer waveguide (active, core and cladding); the spatial mode characteristics are given by the normalized waveguide thickness D [57]:

$$D = \frac{2\pi d}{\lambda} \sqrt{n_{\text{core}}^2 - n_{\text{cladding}}^2} \ , \tag{2.2}$$

where d is the waveguide core thickness, n_{core} is the core index of refraction and n_{cladding} is the cladding index of refraction. For such a waveguide, single mode propagation is obtained for $D < \pi$. By optimizing the width and etching depth of a ridge, single lateral mode emission can be obtained with output powers in the watt range [58]. This concept is utilized in chapter 3, while the influence of the etching depth is shown in Appendix A. Note that in the case of gain guiding, the injected current leads to a small temperature increase that causes an index variation between the waveguide and cladding layers, resulting in a (weak) index guiding effect.

Tapered diode lasers combine a ridge waveguide (RW) section with a TPA section. The RW ensures spatial single mode with a narrow beam waist by means of index guiding, while the TPA section utilizes gain guiding to amplify and guide the emitted light. Such devices provide multiple watts of output power with nearly diffraction limited (single mode) spatial characteristics. As mentioned in the motivation section, TPAs will be utilized in the MOPA systems of this work, see chapter 8.

2.4   Power characteristics

The generated photon emission needs to be amplified by stimulated emission in order to compensate all losses occurring within the resonant cavity. In the case of a Fabry Pérot resonator having a gain medium of length L_{gain} in a cavity of length L_{cav}, positioned between two mirrors of reflectivity R_f and R_b, see Fig. 2.3, the threshold condition is given by [5]

$$\Gamma g_{\text{th}} = \alpha_i + \alpha_{\text{mirror}} = \alpha_i + \frac{1}{2L_{\text{cav}}} \ln\left(\frac{1}{R_f R_b}\right) \ . \tag{2.3}$$

In the above expression, Γg_{th} is the modal gain, given by the product between the threshold material gain g_{th} and the confinement factor Γ. The latter describes the overlap between the optical mode and the active region of the diode laser. α_{mirror} represents the combined photon losses at the diode laser end facets, while α_i describes the internal losses by intrinsic absorption. For a cavity of a certain gain length L_{gain}, the threshold gain can be reduced by either increasing the cavity length L_{cav} or by increasing the reflectivities of the cavity. This being said, increasing the facet reflectivities could lead to catastrophic optical mirror damage (COMD) affecting the performance and lifetime of a laser. Therefore, the front facet is typically made with low reflectivity, in order to reduce the load on the facet due to the out-coupled light.

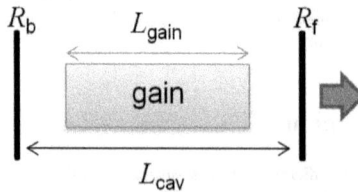

Figure 2.3: *Sketch of a Fabry Pérot resonator.*

After reaching threshold, the output power P shows a linear dependence on the injection current I_{inj} according to [5]

$$P = \eta_i \frac{\alpha_{\text{mirror}}}{\alpha_i + \alpha_{\text{mirror}}} \frac{h\nu}{q} \left(I_{\text{inj}} - I_{\text{th}}\right) = \eta_d \frac{h\nu}{q} \left(I_{\text{inj}} - I_{\text{th}}\right) \ . \tag{2.4}$$

In the above expression, η_i represents the internal efficiency; the fraction of generated photons inside the cavity to the number of carriers, $q = 1.602 \times 10^{-19}$ C is the elementary charge of electrons and I_{th} is threshold current. Finally, η_d is the differential efficiency including the internal efficiency and the different losses. This value is defined as

$$\eta_d = \frac{q}{h\nu} \frac{dP}{dI_{inj}} = \frac{q}{h\nu} S \; , \tag{2.5}$$

where $dP/dI_{inj} = S$ is the slope efficiency measured in W/A.

The third expression that is typically used to describe lasers is the wall-plug efficiency η_c, defined as the ratio between the measured optical power and the injected electrical power:

$$\eta_c = \frac{P}{I_{inj}V} = \frac{\eta_d \, (I_{inj} - I_{th})}{I_{inj} \, (V_d + I_{inj}R_s)} \; . \tag{2.6}$$

In the above expression, V is the voltage drop of the power supply, V_d is the ideal diode voltage (~ 1.2 V for GaAs) and R_s is the series resistance. The wall-plug efficiency can be split into four parts:

$$\eta_c = \underbrace{\eta_i}_{1} \; \underbrace{\frac{\alpha_m}{\alpha_m + \alpha_i}}_{2} \; \underbrace{\frac{h\nu}{q \, (V_d + I_{inj}R_s)}}_{3} \; \underbrace{\frac{I_{inj} - I_{th}}{I_{inj}}}_{4} \; , \tag{2.7}$$

which represents:

1. Internal efficiency > 0.9

2. Mirror loss / threshold gain ~ 0.8

3. Electrical loss ~ 0.9 (series resistance)

4. Excess current ~ 0.9

and thus the product of these factors lies between $\eta_c = 0.5$ and 0.6. Although diode lasers exhibits high wall-plug efficiencies, they are typically limited by thermal effects [59, 60]. As the temperature increases in the active region, the carrier confinement decreases and the internal efficiency decreases [5]. By a temperature change of ΔT, the threshold current I_{th} is changed by:

$$I'_{th} = I_{th} \exp\left(\frac{\Delta T}{T_0}\right) \; , \tag{2.8}$$

where T_0 is the characteristic temperature of the threshold current. Likewise, the differential efficiency η_d is reduced by these thermal effects:

$$\eta'_d = \eta_d \exp\left(\frac{\Delta T}{T_1}\right) \; , \tag{2.9}$$

where T_1 is the characteristic temperature for the above threshold current increment, and is generally two or three times larger than T_0 [52].

These thermal effects are believed to be one of the main reasons behind COMD, which effectively limits the lifetimes of diode lasers [61]. Therefore, heat dissipation of diode lasers is of great importance.

2.5 Spectral properties

The second requirement for lasing is having a resonant optical feedback within the cavity. In the case of a Fabry Pérot resonator of length L_{cav}, standing waves fulfil the requirement:

$$L_{cav} = \frac{m\lambda_0}{2n_{eff}} \ , \tag{2.10}$$

with $m = \mathbb{N}$ being the number of the longitudinal modes, λ_0 is the vacuum wavelength and n_{eff} is the effective refractive index of the waveguide inside the laser [5]. In a similar fashion, the mode spacing $\delta\lambda$, often noted as free spectral range (FSR) is given by

$$\delta\lambda = \frac{\lambda_0^2}{2n_{eff}L_{cav}} \ . \tag{2.11}$$

In a gain medium close to threshold, only the longitudinal modes closest to the maximum modal gain are amplified. In the case of diode lasers, a gain spectrum wider than $\delta\lambda$ enables spectral multimode emission and subsequently enables wavelength tuning [57].

The emission spectra of diode lasers can be narrowed by different means, e.g. by external or internal gratings. Internally, this is done by using DBR or DFB gratings which provide intrinsic wavelength stabilization. In the latter case, the feedback is distributed throughout the device by introducing periodic perturbations of the refractive index along the length of the active region [62]. In both DBR and DFB gratings, the Bragg condition must be fulfilled:

$$2\Lambda \sin(\theta) = m\lambda_B \ , \tag{2.12}$$

where Λ is the grating period, θ is the angle of incidence, $m = 1, 2, 3, ...$ is the order of the Bragg grating and λ_B is the Bragg wavelength [63]. Eq. (2.12) implies that counter propagating waves ($\theta = 90°$) inside the cavity only couple coherently at $\Lambda = m\lambda_B/2$ [57]. Proper engineering of the grating period consequently provides selective feedback, resulting in spectral single-mode emission at λ_B.

2.6 Wavelength tuning

When considering wavelength tuning of diode lasers, three parameters should be examined: the optical gain, the feedback and the mode spacing. This is shown in a generic schematic in Fig. 2.4, representing reflectivity and gain versus wavelength. Note that this is a conceptual sketch which is not to scale, as the gain profile is usually 50 to 100 nm broad, with mode spacings of the order of hundreds of picometers, and with sub-picometer wide reflectivity spectrum.

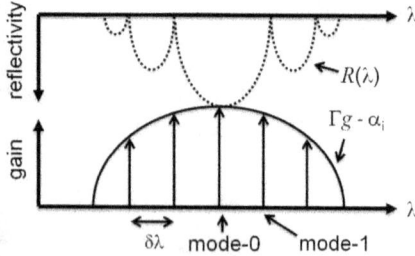

Figure 2.4: *Sketch of the grating reflectivity next to the gain profile including its mode spacing* $\delta\lambda$.

The optical gain in Fig. 2.4 describes the available region of lasing emission and tunability. The shape of the gain profile ($\Gamma g - \alpha_i$) depends on the active medium used for lasing. In the case of quantum well (QW) based diode lasers, the composition and width of the QW structure and the number of embedded QWs determines the gain profile.

The second parameter, the optical feedback, and in particular the reflectivity of the grating defines the part of the gain that sees feedback.

The final parameter is the mode spacing, which is represented as vertical arrows within the shape of the gain profile in Fig. 2.4. Having a single longitudinal mode fit within the reflectivity curve $R(\lambda)$ will result in emission of a single mode. When multiple modes fits, fully or partially, then the laser is characterized by its side-mode suppression ratio (SMSR). This is defined as the power ratio between the main mode λ_0, and the first higher-order mode λ_1:

$$ SMSR = \frac{P(\lambda_0)}{P(\lambda_1)} \; . \tag{2.13} $$

Wavelength tuning can be obtained by different means, which can be described by the movement of the curves in Fig. 2.4 relative to each other. By temperature tuning of the whole device, e.g. changing the heat sink temperature, the reflectivity curve $R(\lambda)$ and the gain profile will shift together, resulting in a so-called continuous wavelength tuning. In this case, mode-0 is maintained as the lasing mode. When heating the grating section only, lasing will start at mode-0 which will tune a distance smaller than the mode spacing $\delta\lambda$, until its neighbouring peak (mode-1), fits better within the moving reflectivity curve. This will cause a quasi-continuous wavelength tuning, where the wavelength will jump from one mode to another, causing a step-like behaviour. This is described as mode jump or mode hop which means that a small wavelength region between the modes will not be covered by the wavelength tuning.

Alternatively, a three section laser with a gain, phase and a grating section can provide continuous wavelength tuning. Tuning of the grating section will shift the reflectivity curve, while a control of a passive phase section can adjust the position of the individual modes accordingly, thus ensuring that the mode-0 will remain as the lasing mode.

Tuning of the grating section can be done either electrically or thermally, which in both cases causes a slight change in the index of refraction of the grating structure and thus shifts the Bragg wavelength, see eq. (2.12). In the case of electrical tuning, a current is injected into the grating section causing a decrease in the index of refraction, see Fig. 2.5(a). Electrical tuning is well established in InP based diode lasers emitting around 1.3 and at 1.55 μm and is used for optical communication. In the case of GaAs based lasers, electrical tuning remains a challenge due to low non-radiative carrier lifetime (< 5 ns) [64]. As a result, thermal tuning is preferred in GaAs lasers where resistive heater elements provide an increase in the index of refraction with increased temperature, see 2.5(b). By increasing the heater current, the index of refraction increases causing an increases the effective index of refraction n_{eff}, which shifts the grating wavelength. The applied heater elements of this work will be described in chapter 6.

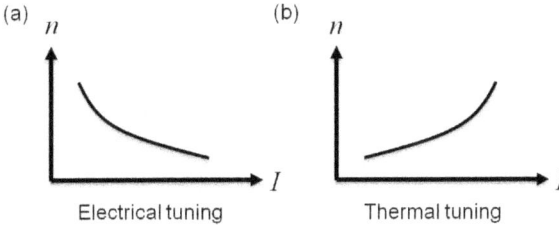

Figure 2.5: *Sketch of index of refraction n during (a) electrical tuning and (b) thermal tuning as function of tuning current.*

Finally, DFB lasers can be tuned electrically by varying the injection current, however they provide a reduced tuning range (in comparison to tunable DBR-lasers), with mode hops and output power modulation during the wavelength tuning [41] .

Chapter 3

DBR-RW lasers

GaAs based semiconductor lasers with DBR and DFB gratings can provide diffraction limited and tunable narrow linewidth emission in the near IR, but emit an output power below the watt level [40, 41]. The excellent beam quality of these light sources however, makes them ideal in MOPA configurations where the light can be tightly focused, and thus efficiently coupled into a PA. In this work, DBR based lasers were chosen over DFB lasers, as the latter are usually tuned by varying the injection current. This results in a power and beam quality variation during the wavelength tuning that are undesired in MOPA configurations. Finally, grating degradation is expected over the lifetime of a DFB laser, due to the injection current sent through the DFB grating.

As described in the introduction chapter, the first concept of tunable lasers uses multi-wavelength (multi-arm) DBR-RW lasers in MOPA configurations. These systems require powers above 15 mW as this is the typical saturation power of TPAs [21]. The multi-wavelength approach is obtained by monolithically combining individual laser arms with corresponding DBR gratings into a common laser aperture. The motivation is then to distance each DBR grating (spectrally) so that in combination, a wider tuning range is obtained compared to that of a "single-arm" tunable DBR-RW laser.

In this chapter, the developed light sources will be described, including its material data parameters, vertical and lateral structures, as well as the device mounting and facet coating. In addition, two reference lasers will be characterized and their results will be used when comparing the more sophisticated lasers of chapter 4 and 5. This characterization focuses on the electro-optical performance, wavelength stabilization and spatial characteristics, with the aim on implementation into MOPA systems. The tuning performance of these lasers will be described in chapter 6.

3.1 Vertical structure

The vertical structure used for the DBR-RW based tunable lasers of this and the following two chapters will be described next.

The vertical structure is a large optical-cavity (LOC) [65, 66], with an InGaAs SQW as the active layer, see Fig. 3.1. The 7 nm thick SQW is sandwiched between GaAsP spacer layers, and is index guided by AlGaAs waveguide and cladding layers. This vertical structure is grown using metal organic vapour phase epitaxy (MOVPE). The developed structure emits transverse electrical (TE) polarized light, and has a vertical far field angle of 26.7° measured at full width at half maximum (FWHM), see Fig. 3.2.

3.1.1 Internal material parameters

For reference purposes, the developed vertical structure was characterized by measuring the threshold current and the slope efficiency of an uncoated 1 mm long Fabry-Pérot laser with a stripe width of 100 μm. Under pulsed mode; 1 μs long current pulses at a repetition rate of 1 kHz, a threshold current $I_{th} = 158$ mA, a slope efficiency $S = 0.60$ W/A and a differential efficiency of $\eta_D = 0.94$ were determined. Under the same excitation conditions, the figures of merit for the structure were obtained from power-current characteristics for devices with 100 μm stripe widths and 9 different resonator/cavity lengths:

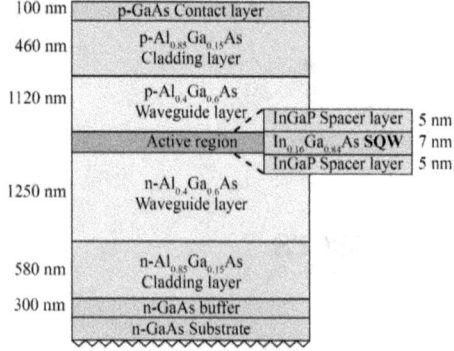

100 nm	p-GaAs Contact layer
460 nm	p-Al$_{0.85}$Ga$_{0.15}$As Cladding layer
1120 nm	p-Al$_{0.4}$Ga$_{0.6}$As Waveguide layer
	Active region
1250 nm	n-Al$_{0.4}$Ga$_{0.6}$As Waveguide layer
580 nm	n-Al$_{0.85}$Ga$_{0.15}$As Cladding layer
300 nm	n-GaAs buffer
	n-GaAs Substrate

InGaP Spacer layer — 5 nm
In$_{0.16}$Ga$_{0.84}$As SQW — 7 nm
InGaP Spacer layer — 5 nm

Figure 3.1: *LOC vertical structure of the studied lasers in chapter 3 to 6.*

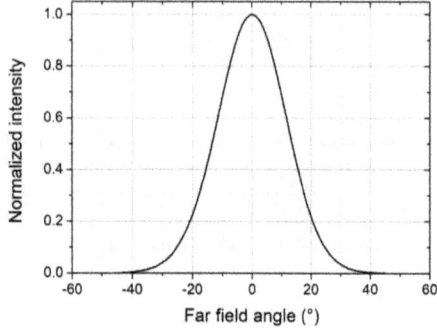

Figure 3.2: *Measured vertical far field profile of the developed LOC structure.*

$L_{\text{cav}} = [400, 600, 800, \dots, 1800, 2000]$ μm. From the obtained threshold current I_{th} and slope efficiency S values at each resonator length, the figures of merit given in Table 3.1 were determined. This model assumes a logarithmic dependence of the modal gain on the current density [52], and is described further in Appendix B. Using this vertical structure with a SQW provides a high internal quantum efficiency and modal gain, together with low internal losses and a good temperature stability. This makes this vertical structure well-suited for laser processing.

Table 3.1: *Material data parameters of the developed vertical structure.*

Parameter	Symbol	Value
Internal quantum efficiency	η_{i}	≈ 0.95
Internal losses	α_{i}	1.2 cm^{-1}
Transparency current density	J_{TR}	72.5 A/cm^2
Threshold current density for devices with an infinite length	J_{∞}	78.2 A/cm^2
Modal gain coefficient	Γg_0	15.9 cm^{-1}
Characteristic temperature of the threshold current	T_0	133 K

14

3.2 Lateral structure

The RW structures of the DBR-RW lasers are 2.5 μm wide, made using projection lithography with an I-line wafer stepper. These RW structures are deeply etched (down to $\sim$ 1372 nm), and a comparison between deeply and flat etched RW structures can be found in Appendix A.

The DBR gratings are 7th order deeply etched surface gratings made with E-beam lithography. Fig. 3.3 shows a scanning electron microscope (SEM) image of the developed grating structure.

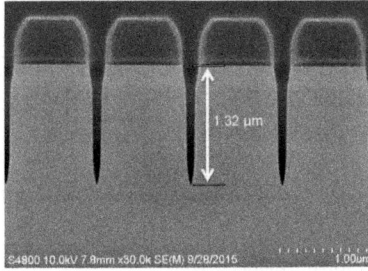

Figure 3.3: *SEM image of the 7th order DBR grating.*

The facets of the investigated devices are cleaned by applying atomic hydrogen and a sealing process with ZnSe before being reflectivity coated, according to the technology described by Ressel et al. [67]. All lasers in chapter 3 to chapter 6 have a front facet reflectivity of $R_f \approx 5\%$ and an AR coating on the back facets, with a reflectivity about the order of $R_b \approx 5 \times 10^{-5}$, see Fig. 3.4. The back facets are AR coated to reduce optical feedback effects of back reflected waves that do not propagate inside the waveguide.

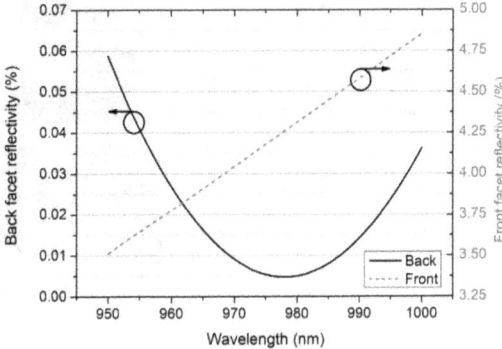

Figure 3.4: *Measured reflectivity curves of the back (left) and front (right) facets at different wavelengths.*

3.3 Device mounting

All the investigated lasers were soldered p-side up on individual CuW sub-mounts using AuSn. These were then soldered on AlN heat spreaders which then were soldered on conduction cooled package (CCP) using PbSn. The p-side contacts were realized by wire bonding.

3.4 Reference lasers

In the following, a "straight" and an S-bend shaped DBR-RW laser will be considered and their characteristics will be used as reference, especially when considering the multi-arm lasers of the next chapters.
The "straight" or regular DBR-RW laser has a 1000 μm long DBR grating, a 2000 μm long gain section
and a RW width of 2.5 μm, see Fig. 3.5(a). The S-bend reference laser has a 1000 μm long DBR grating,
a 3000 μm long gain section and a RW width of 2.5 μm, see Fig. 3.5(b). The S-bend shape follows a
Co-sinusoidal form which will be described in chapter 4.

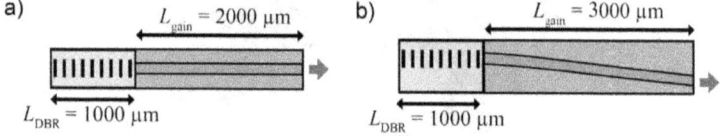

Figure 3.5: *Sketch of the investigated (a) straight and (b) S-bend DBR-RW reference lasers.*

3.4.1 Electro-optical characteristics

The power-voltage-current (PUI) characteristics of the two reference lasers at a heat sink temperature
of $T = 25°C$ are shown in Fig. 3.6(a). Each laser was measured for $I_{inj} = [0, 500]$ mA with a step size
of $\Delta I_{inj} = 5$ mA. The straight reference laser has a threshold current of $I_{th} = 30$ mA, a slope efficiency
$S = 0.84$ W/A and an output power $P_{max} = 365$ mW at $I_{inj} = 500$ mA. For this laser, strong kinks can
be seen at higher output powers.

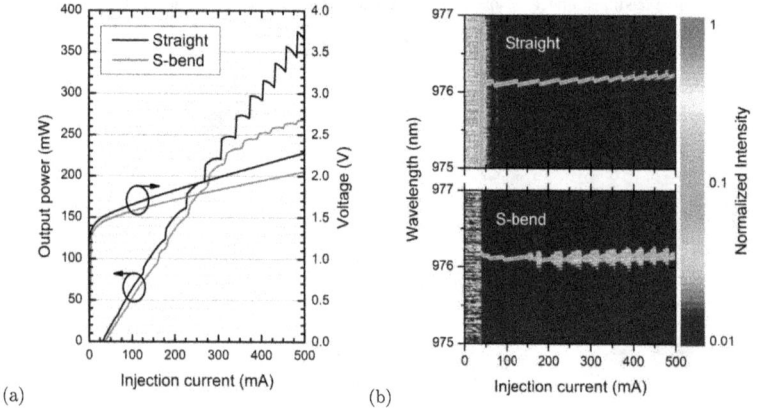

Figure 3.6: *Electro-optical characteristics of the DBR-RW straight and the S-bend reference lasers at $T =$
25°C showing (a) the PUI curves and (b) the corresponding emission wavelengths at different injection
currents.*

The S-bend reference laser has a threshold current of $I_{th} = 35$ mA, a slope efficiency $S = 0.81$ W/A and
an output power $P_{max} = 268$ mW at $I_{inj} = 500$ mA. The slope efficiencies of both reference lasers were
obtained from linear fits between $I_{inj} = [50, 300]$ mA, a region above threshold and below the role-over
effect. This measurement indicates that the S-bend structure introduces additional (bending) losses,
which can be seen in the reduced output power and the earlier role-over effect. The longer gain section
of the S-bend structure introduces more absorption, however these loses are expected to be smaller than
the bending losses.

Likewise, the spectral behaviour of the reference lasers was obtained and is shown in Fig. 3.6(b). These plots show false colour contour diagrams, obtained using a double echelle monochromator *DEMON* from LTB Lasertechnik Berlin, with a spectral resolution of 17 pm at 976 nm. The straight reference laser emits around 976.1 nm, which is single-mode over the entire investigated injection current range. This laser has a mode spacing of about $\delta\lambda = 64$ pm, providing an effective cavity length of about $L_{cav} = 2.07$ mm for an effective index of $n_{eff} = 3.66$, see eq. (2.11). This is in agreement with the manufactured gain length of 2 mm, plus a small penetration depth into the grating section. The effect of penetration depth is described in Appendix C.

The S-bend reference laser emits at similar wavelengths, however showing regions of multi-mode operation appearing between the individual mode-jumps. The mode spacing of this laser is about $\delta\lambda = 42$ pm, providing an effective cavity length of about $L_{cav} = 3.15$ mm, which is close to the 3 mm long gain section of this laser, see Fig. 3.5(b).

Finally, Fig. 3.7 shows the emission spectra at maximum output powers of both reference lasers, indicating an emission width below 17 pm (resolution limited)

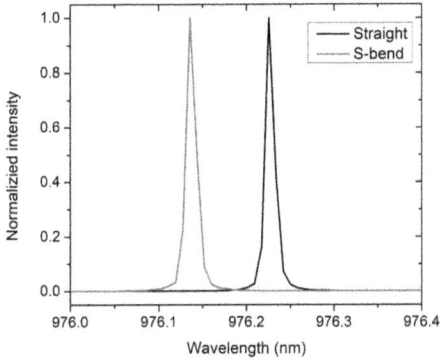

Figure 3.7: *Individual optical spectra of the two reference lasers at* $I_{inj} = 500$ mA *and* $T = 25°C$.

3.4.2 Spatial characteristics

The spatial characteristics of the two reference lasers were obtained according to the moving slit method [68]. The normalized lateral intensity distributions of the near and far fields are shown side-by-side in Fig. 3.8 at different injection currents. Gaussian shaped near and far fields profiles are observed for both lasers which stay unchanged at increasing output powers. A beam waist of about 3.9 μm, a far field angle of 17.8° and an $M^2 = 1.1$ (obtained at $1/e^2$) were measured, remaining more or less unchanged (within the measurement accuracy) at increasing output powers.

The close resemblance between the two reference lasers suggest that the S-bend structure does not change the spatial characteristics, as those are mainly defined by the RW width and laser aperture, which are similar for both lasers.

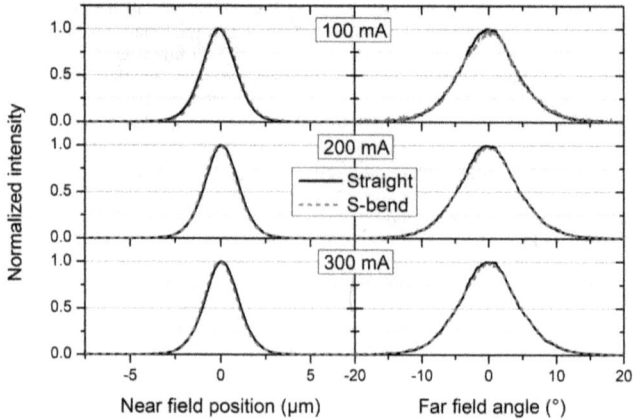

Figure 3.8: *Normalized near field positions (left) and far field profiles (right) of the straight and S-bend reference lasers at different injection currents at* $T = 25°C$.

3.4.3 Summary

The DBR-RW reference laser showed overall expected behaviour with up to 365 mW of single-mode emission. The S-bend reference laser showed a reduced output power of up to 268 mW, which together with the earlier role-over effect observed suggest additional losses due to the S-bend shape.

Spatially, both lasers showed Gaussian-like near and far field profiles, which stayed unchanged with increasing output powers. These results suggest that the S-bend structure does not affect the spatial characteristics, and these results will serve as reference when considering the more sophisticated laser structures of chapter 4 and 5.

Chapter 4

Y-branch DBR-RW lasers

Y-branch DBR-RW lasers combine two individual DBR-RW structures in a Y-shaped intersection to provide emission at two wavelengths from the same aperature. These light sources are suitable for beat signal generation [69, 70], terahertz frequency generation [20, 71] and within the field of shifted excitation Raman spectroscopy (SERDS) [72, 73]. Y-branch lasers at 671 nm and 785 nm have recently been demonstrated with output powers of 100 mW and 200 mW, respectively [74, 75].

In this work, 3 different types of Y-branch DBR-RW lasers will be considered, where two S-bend structures are combined into a common front section and integrated into an active cavity. The DBR gratings of each branch/arm have different periods. Thus, depending on how the arms are electrically driven, the Y-branch lasers can be either switched from one wavelength to the other, or can operate at two wavelengths simultaneously. The contact scheme for the Y-branch lasers is shown in Fig. 4.2(a), consisting of two arms, that are combined in a Y (common) section, before being emitted from the front section. These different sections (arms, Y- and front section) have isolated gold surface contacts which enables separate/individual control of the current injection. This being said, in the first part of this investigation (Sec. 4.1 and 4.2), all three sections are controlled in a parallel connection; Left- or right arm + Y + front sections.

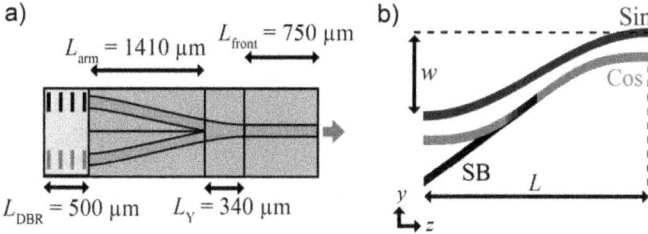

Figure 4.1: *Sketch of (a) the contact scheme and (b) the three implemented S-bends structures of the investigated Y-branch lasers.*

The three Y-branch lasers differ by having a sinusoidal (Sin), a co-sinusoidal (Cos), and a Cos based single bend (SB) waveguide curvatures, see Fig. 4.1(b). The lateral position of the Sin and Cos S-bend structures were generated using the following formulas [76]:

$$Y_{\mathrm{Sin}}(z) = \frac{w}{L}z - \frac{w}{2\pi}\sin\left(\frac{2\pi z}{L}\right) \ , \tag{4.1}$$

$$Y_{\mathrm{Cos}}(z) = \frac{w}{2}\left[1 - \cos\left(\frac{\pi z}{L}\right)\right] \ , \tag{4.2}$$

where $w = 40\ \mu$m is the lateral offset for a bending length of $L = 2000\ \mu$m, see Fig. 4.1(b). After the merge of the two branches, the light is guided through a 500 μm long straight front waveguide section.

The cos based SB is basically a Cos bend, consisting of an angled (1.8°) RW section, which is bent only once (thus noted SB) using a Cos bend ($w = 40$ μm, $L = 2000$ μm). Likewise, its two arms are combined and emitted from a 500 μm long front section. Thus it follows the Cos bend formula but only between $z = [L/2, L]$ and has a lateral offset of 67 μm. In comparison, the Sin and Cos each have two bends (forming an S-shape) providing a lateral offset with a smooth curvature, see Fig. 4.1(b). The angle value of 1.8° was chosen as it provides a smooth transition between the the angled and the Cos SB section. In addition, the DBR gratings of this laser are also angled (±1.8°) relative to front section and to the crystal axis.

The complete RW structures of the three investigated Y-branch lasers are shown in Fig. 4.2. As can be seen, the Sin and Cos lasers only differs in the S-bend structure, where the Sin S-bend has a bending radius of 15.9 mm while the Cos has a bending radius of 20.2 mm. The SB has a similar bending radius as the Cos S-bend. The arms of the Sin based Y-branch laser have an arc length of 2500.6 μm, the Cos have a length of 2500.5 μm, while the arms of the SB have a length of 2500.3 μm. In all three cases relatively large bend radii, obtained by setting $w \ll L$, were used to reduce bend losses according to the etch depth and effective index step methods [77, 78].

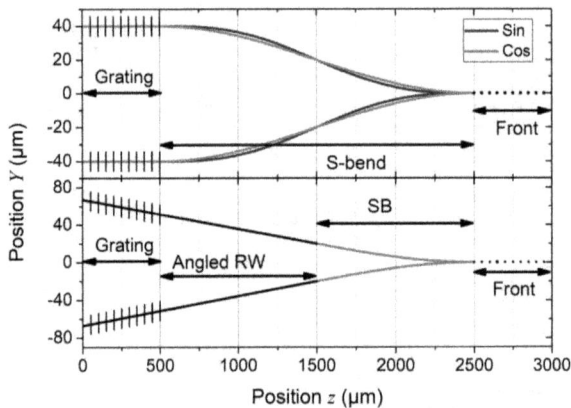

Figure 4.2: *Sketch of three investigated Y-branch DBR-RW lasers.*

4.1 Electro-optical characteristics

Similarly to the reference lasers, the three Y-branch lasers were characterized at a heat sink temperature of $T = 25°$C. The power characteristics of these lasers, for both the right and left arms are shown in Fig. 4.3. These measurements were made for injection currents between $I_{inj} = [0, 500]$ mA with a step size of $\Delta I_{inj} = 5$ mA.

The Y-branch lasers provide comparable output powers from the left and right arm, with some differences at high output powers. A threshold current of about $I_{th} = 35$ mA was observed for both arms of all three lasers. At an injection current of $I_{inj} = 500$ mA, the Y-branch lasers provide output powers of about $P_{Sin/Cos/SB} = [237/253/253]$ mW, respectively. Linear fits between $I_{inj} = [50, 300]$ mA, a range which is above threshold and below the roll-over effect, provide slope efficiencies of about $S_{Sin/Cos/SB} = [0.70/0.69/0.69]$ W/A, for the Sin, Cos, and the SB lasers, respectively. As indicated in Fig. 4.3, the roll-over effect can be estimated at an injection current of $I_{inj,Sin/Cos/SB} = [300/355/310]$ mA for the Y-branch lasers, respectively. Kinks can be seen at higher output powers for all three Y-branch laser, which was also observed from the two reference lasers, see Fig. 3.6(a).

The three lasers have similar threshold currents that are comparable to the two reference lasers which had 30 and 35 mA, respectively (see Sec. 3.4). This being said, they achieve less maximum output powers

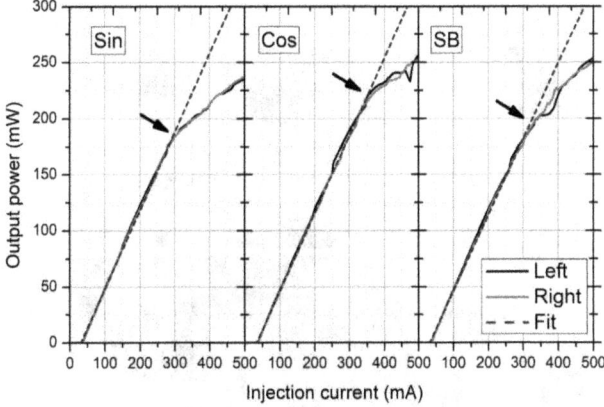

Figure 4.3: *Power characteristics of the investigated Y-branch lasers at $T = 25°C$ including linear fits.*

when compared to the straight DBR-RW reference laser which had $P_{max} = 365$ mW. The Y-branch lasers emit comparable output powers to the S-bend reference laser with $P_{max} = 268$ mW. Finally, the slope efficiencies of the Y-branch lasers are smaller than the 0.84 and 0.81 W/A obtained for the straight and the S-bend reference lasers, respectively. This is perhaps not surprising as the intersection between the two arms in the Y-branch lasers is expected to provide additional losses, which will be considered later in this chapter.

The corresponding spectral behaviour of the Y-branch lasers is shown in Fig. 4.4. The right arm of the Sin Y-branch laser emits at wavelength of about $\lambda_{Sin,right} = 976.0$ nm while the left arm emits at $\lambda_{Sin,left} = 973.7$ nm. Both of the Sin arms are single-mode until $I_{inj} = 375$ mA, at which multi-mode operation take place. The effect of multi-mode operation is also noticeable in the power characteristics in Fig. 4.3, where sudden drop and/or kinks starts to appear.

The Cos Y-branch laser emits at $\lambda_{Cos,right} = 975.9$ nm and at $\lambda_{Cos,left} = 973.6$ nm, for the right and left arm, respectively. The right arm shows multi-mode operation at $I_{inj} = 435$ mA, while the left arm already becomes multi-mode at $I_{inj} = 400$ mA. This difference is also visible in Fig. 4.3, as the output power of the right and the left arms separate around $I_{inj} = 405$ mA.

Finally, the SB Y-branch laser emits at $\lambda_{SB,right} = 975.9$ nm and at $\lambda_{SB,left} = 973.6$ nm. The right arm becomes multi-mode at $I_{inj} = 360$ mA, while the left arm becomes multi-mode at $I_{inj} = 310$ mA. Once again, in agreement with the power curves in Fig. 4.3.

By investigating the optical mappings in Fig. 4.4, the mode spacing of the Y-branch lasers is estimated to be $\delta\lambda = 52$ pm. This corresponds to an effective cavity length of $L_{eff,cav} = 2.54$ mm, in agreement with the manufactured cavity lengths of ~ 2500 mm, see Fig. 4.2. The additional 40 μm corresponds to a penetration into the grating section.

The individual optical spectra at maximum single-mode output powers are shown in Fig. 4.5. From this measurement, the emission linewidth of all three Y-branch lasers is estimated to be smaller than 17 pm, limited by the spectrometer. The spectral distance between the right and left arms of the Y-branch lasers was manufactured to be 2.3 nm. Note also that each pairs of wavelengths were designed similarly and thus the wavelength mismatch between the three lasers is due to manufacturing tolerances.

The three Y-branch lasers show single-mode operation until a "breaking" point where they become multi-mode. This is different to the spectral behaviour of the straight reference laser, which maintained single-mode operation over the entire investigated injection current range. A difference is also observed when comparing the Y-branch lasers to the S-bend reference laser which only showed multi-mode operation at the mode jumps, see Fig. 3.6(b). Overall, the electro-optical measurements show a clear difference between the reference and the Y-branch lasers, and suggests different guiding and/or loss mechanisms.

This will be investigated further in this chapter.

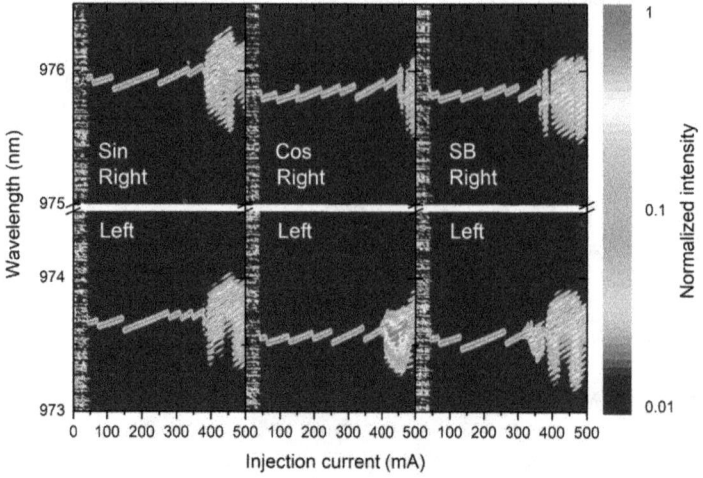

Figure 4.4: *False color contour plots of the emission wavelength of the studied lasers, as function of the injection current at $T = 25°C$.*

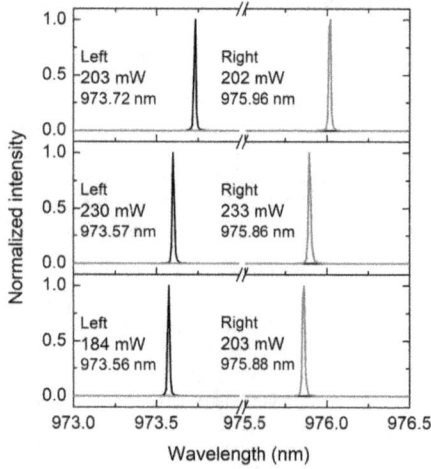

Figure 4.5: *Individual optical spectra of the Y-branch lasers at maximum single-mode output powers at $T = 25°C$.*

4.2 Spatial characteristics

The near field profiles of the three Y-branch lasers are shown side-by-side at different injection currents in Fig. 4.6. A Gaussian shaped near field profile is observed for the Y-branch lasers, similarly to the reference lasers (see Fig. 3.8), however with side-modes (tails) at low output powers. These side-modes have mirror symmetry between the right and the left arms, and are suppressed at higher output powers. The right arm of the Sin laser has its side-mode on the left side (negative position). The right arms of both the Cos and the SB have their side-modes on the right side (positive position), and vice versa for the left arms. This is due to the different curvature shapes of the Sin and Cos S-bends, as will be shown in the numerical simulations in section 4.4. The SB follows the curvature of the second Cos bend and hence have similar behaviour. The observed side-modes are mostly evident at low powers, and are suppressed at higher output powers for all three lasers. Note that at $I_{inj} = 500$ mA, corresponding to spectral multi-mode operation (see Fig. 4.4), the beam profiles are not changed much.

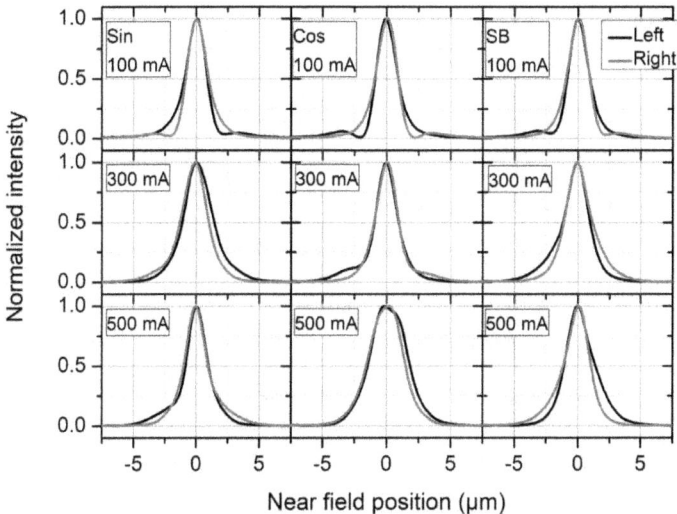

Figure 4.6: *Normalized near field profiles of the investigated lasers, at different injection currents at* $T = 25°C$.

Likewise, the far field profiles are shown side-by-side in Fig. 4.7 for the three Y-branch lasers. At low output powers, the far field profiles of the Y-branch lasers consist of multiple high intensity peaks. This effect is however reduced at higher output powers, and a single peak is measured with multiple low intensity neighbouring peaks. By comparing the three S-bend shapes, it seems that the Sin shows least changes in the far field profile with increasing output powers and the one being the most Gaussian-like. Similar to the near field profiles, mirror symmetry is observed between the far field profiles.

The Cos based laser has two major peaks which merge at higher output powers. In comparison to [75] which reports on a Cos based S-bend Y-branch laser at 785 nm, the far field profiles seems to have the opposite effect of separating at higher output powers. This difference is believed to be caused by the different polarization states of the two lasers; TE versus transverse magnetic (TM). In addition, the different material data parameters, in particular the transparency current density, which for the 785 nm Y-branch laser was $J_{TR} = 135$ A/cm^2 while for this investigation is $J_{TR} = 72.5$ A/cm^2, see Table 3.1.

The main peaks of the left and right arms of the SB Y-branch laser overlap at low output powers, and separates at higher output powers.

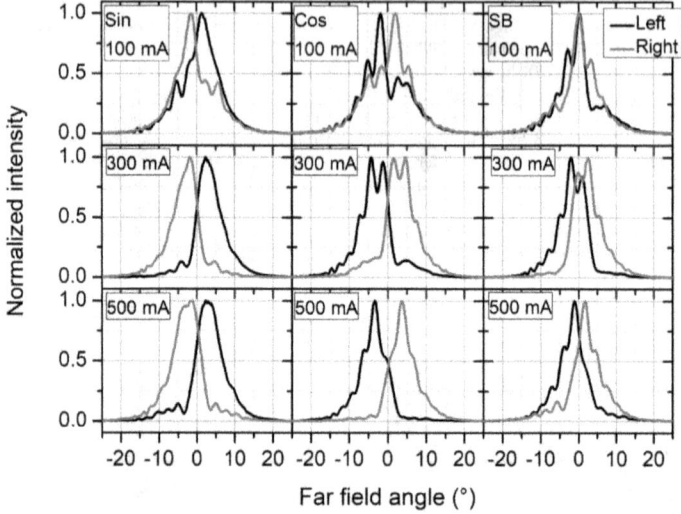

Figure 4.7: *Normalized far field profiles of the investigated lasers, at different injection currents at* $T = 25°C$.

The results of the spatial characteristics are summarized in Table 4.1 for both the reference and the Y-branch lasers. It includes the near field widths, the far field angles and the propagation factors $M^2_{2.\text{Mom.}}$ of the slow axis at the corresponding injection currents.

In comparison to the reference lasers, the spatial characteristics of the Y-branch lasers changes with increasing output powers. It is seen that the propagation factors $M^2_{2.\text{Mom.}}$ improves with increasing output powers, and it is also observed that the width of the near fields decreases at higher powers. The biggest influence on the beam quality is in the far field, which changes the most with increasing output powers. This is in comparison to the near fields which are more comparable at different output powers. The overall far field behaviour is ambiguous and will be considered in the next sections.

Table 4.1: *Spatial characteristics of the reference and the Y-branch lasers at different injection currents.*

Laser at I_{Inj} [mA]	Near field width [μm]	Far field angle [°]	Beam quality factor $M^2_{2.\text{Mom.}}$
Straight DBR: 100 / 200 / 300	4.1 / 4.0 / 4.1	18.9 / 19.1 / 19.1	1.1 / 1.1 / 1.1
S-bend DBR: 100 / 300 / 500	4.2 / 4.1 / 4.2	19.1 / 19.6 / 19.6	1.1 / 1.1 / 1.1
Sin left: 100 / 300 / 500	6.2 / 6.0 / 5.9	22.0 / 17.5 / 19.4	1.9 / 1.5 / 1.6
Sin right: 100 / 300 / 500	6.3 / 5.9 / 5.9	23.0 / 18.8 / 20.4	2.1 / 1.6 / 1.7
Cos left: 100 / 300 / 500	7.4 / 6.5 / 6.4	24.5 / 20.6 / 17.8	2.6 / 1.9 / 1.6
Cos right: 100 / 300 / 500	7.1 / 5.9 / 6.2	23.8 / 19.9 / 17.8	2.4 / 1.7 / 1.6
SB left: 100 / 300 / 500	7.0 / 6.4 / 6.0	23.1 / 17.8 / 20.5	2.3 / 1.6 / 1.7
SB right: 100 / 300 / 500	6.7 / 6.1 / 5.8	22.3 / 17.6 / 20.6	2.1 / 1.5 / 1.7

The results in Fig. 4.7 and in Table 4.1 suggest that the Sin S-bend has the most "Gaussian-like" shape, and that its spatial characteristics changes the least with increasing output powers. This being said, all three Y-branch lasers exhibits different spatial characteristics when compared to the reference lasers. The reason for this will be considered next.

4.3 Simulations of single S-bend structures

To understand the spatial behaviour observed so far, passive waveguide simulations were performed on the investigated structures and compared to the experimental results. These simulations were done with the commercial *FIMMWAVE* software [79], based on a complex effective index solver. Note that these passive RW simulations are not expected to fully agree with the experimental results obtained from active laser devices. Nonetheless, it is meaningful to observe light propagation through such passive structures.

In this section, simulations are carried out on single S-bend structure; i.e. single Sin, Cos and SB RW structures, which are then compared to the S-bend (Cos based) reference laser. In addition, each simulation is made with/without a (500 μm long) straight front section, in order to investigate the influence of the front section on the beam quality, see Fig. 4.8. The simulated results next to the measured near field profile of the S-bend reference laser (at $I_{\text{inj}} = 100$ mA) are shown in Fig. 4.9. Note that this low injection current was chosen as it resembles a passive RW case the most.

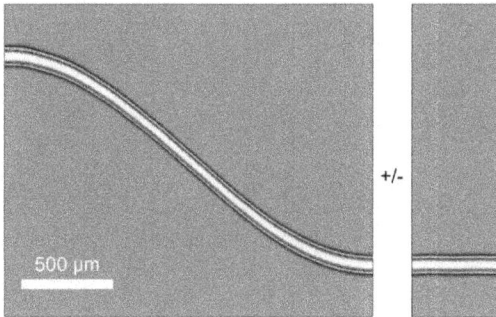

Figure 4.8: *Simulation of light propagation through a passive Cos based RW S-bend, with or without a 500 μm long straight front section. The brighter the color the stronger the field intensity.*

From Fig. 4.9 it is seen that the four cases show similar "Gaussian-like" near fields, with a beam waist of about 3.4 μm (measured at $1/e^2$). This is also maintained regardless of having a front section or not. This is as expected as all the investigated structures have a small lateral offset compared to the bending length: $w \ll L$, and all share the same RW width.

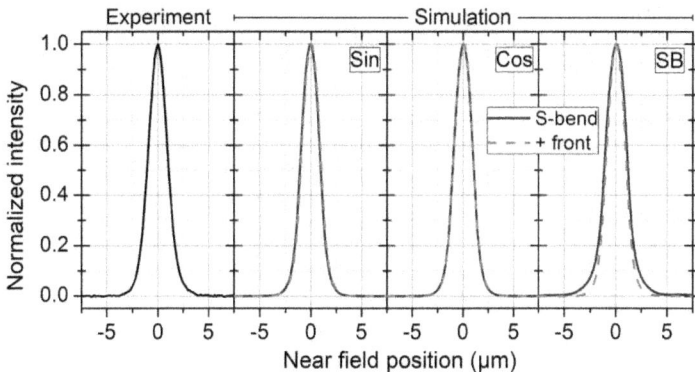

Figure 4.9: *Measured near field position of the S-bend reference laser, next to simulated near fields of the different S-bend structures, with (dashed) or without (solid) a 500 μm long straight front section.*

In a similar fashion, simulations of the far field profiles of the single S-bend structure were carried out. Fig. 4.10(top) shows the measured far field of the S-bend reference laser next to the simulated fields of the S-bend structures. Fig. 4.10(bottom) shows the corresponding simulated fields of the S-bends with a front section. A far field angle (at $1/e^2$) of $17.7°$ was obtained for the experimental measurement, while the simulated profiles have values between $19.0°$ and $18.4°$. In the case of the simulated fields, a modulation effect is seen for all six cases, however an overall Gaussian-like shape is maintained and no additional high intensity peaks are observed.

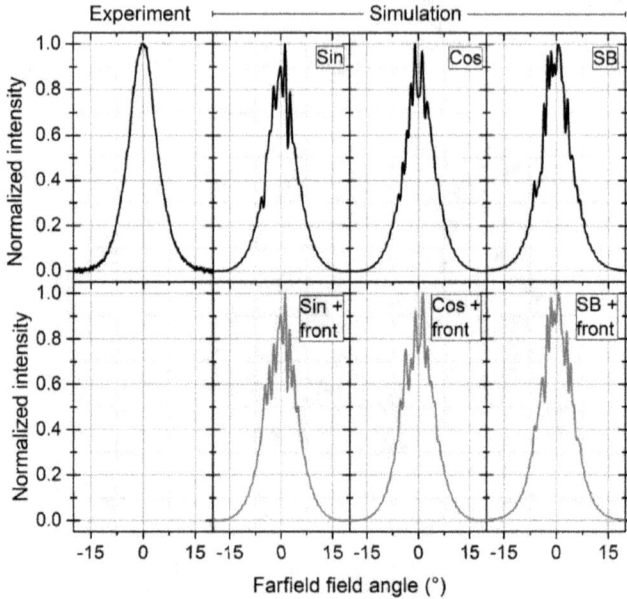

Figure 4.10: *Top: Measured far field profile of the S-bend reference laser, next to simulated far field profiles of the different S-bend structures. Bottom: Simulated far field profiles of the S-bend structures with a front section.*

The simulated results of this section indicate that passive S-bend structures provide a modulation effect on the far field profiles. This being said, the worsen beam quality of the Y-branch lasers, in particular the multiple far field peaks is less dependent on the S-bend structures. The same can be said about the front section, as its introduction did not influence the far field shape (in all three S-bend cases). This suggests that the beam quality is more dependent on the combination of two S-bends; the Y-section rather than the S-bend shapes or the front section.

4.4 Simulations of Y-branch structures

The results of the previous section suggest that the spatial characteristics of the Y-branch lasers are less dependent on the S-bend shape nor the front section. Therefore, this section continues the numerical investigations by studying passive Y-branch structures. Fig. 4.11 shows the simulated propagation profiles, from left to right, for the case of (a) a "straight" RW, (b) Sin, (c) Cos, and (d) SB based Y-branch RW structures.

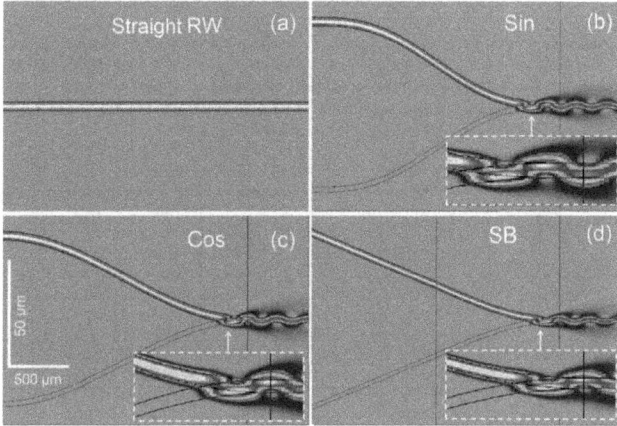

Figure 4.11: *Simulation of light propagation through passive RW structures with similar dimensions to the developed active devices.*

In the case of the "straight" RW, the light is transmitted uniformly through the waveguide. For the Y-branch structures however, an abrupt change occurs at the intersection point in the Y-section. At this point, the waveguide suddenly becomes twice as wide, see the inset zooms in Fig. 4.11(b)-(d). From these simulations, the near field profiles (of the left arms) at the front facet were obtained. These are shown in Fig. 4.12 together with the corresponding experimental results (at low output power) of the Y-branch DBR lasers.

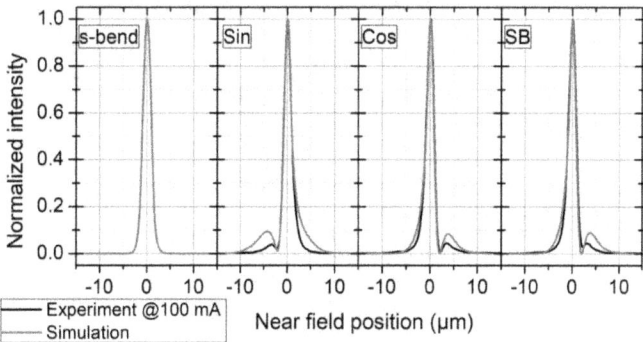

Figure 4.12: *Simulated near field positions of the passive RW structures together with the experimentally obtained ones.*

The near field profile of the RW structure shows a Gaussian-like profile (similar to the S-bend structures), which overlap with the measured near field of the straight reference laser. Likewise, the Y-branch passive RW structures show a "relatively" good agreement with the Y-branch lasers. The mismatch between the measured and simulated near fields can perhaps be reduced by finding a proper effective index Δn_{eff} value used in the simulation (by means of fitting). However, this was not done as it will not bring further understanding to this investigation. As mentioned earlier, these passive RW simulations are not expected to fully agree with the measured results.

The simulated results have overall similar shape to the measured ones, and show opposite side-mode position between the Sin and the Cos/SB cases, in agreement with the experimental results. This being said, and as indicated in the insets of Fig. 4.11(b)-(d), the position of the side peak depends on where the light is measured from (as it zig-zags after the Y-section), i.e. how long the front section is. This is simulated in Fig. 4.13, which shows the near field positions of a Sin/Cos based passive Y-branch structures, having a 30, 85 and 120 μm long front sections. The neighbouring peak position oscillates with a period of about ≈ 180 μm as can be estimated from Fig. 4.11.

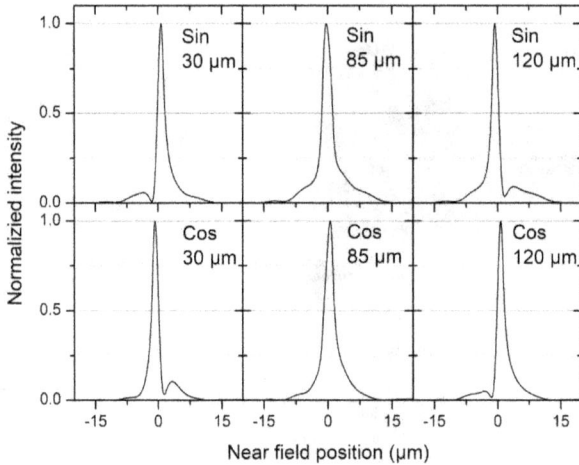

Figure 4.13: *Simulated near field positions of the Sin and Cos based Y-branch structures with different front section lengths.*

The far field distributions were also simulated and compared to the experimental results, see Fig. 4.14. Once again, a good match is seen between the passive RW structure and the reference laser, as both show a Gaussian profile with similar form and width. In the case of the three Y-branch passive RW structures however, the simulated results show two main peaks, while the experimental laser results shows multiple peaks. The simulated results reproduce to a certain limit the experimental results, as the simulated two peaks of the Sin and Cos structures are also observed in the experimental measurements. Some of the smaller peaks are however missing. The main difference is seen for the SB Y-branch structure where the simulated and measured results differ the most. As mentioned earlier, the applied RW simulations have obvious limitations as clearly seen in Fig. 4.14. Nonetheless, the overall profile; particularly the width of the main peak (at FWHM) seems to somewhat match for the experimental and simulated cases. The simulated results provide some components of the measured far field profiles, although a considerable part is missing.

Next, transmission coefficients through each of the simulated RW structures were obtained from the *Fimmwave* software by calculating the S-matrices [80]. This was done for 2500 μm long RW structure with a $\alpha_i = 1.2$ cm^{-1}. A convergence test of the transmission coefficients as function of the applied grid size (resolution) used in these simulations can be found in Appendix D.

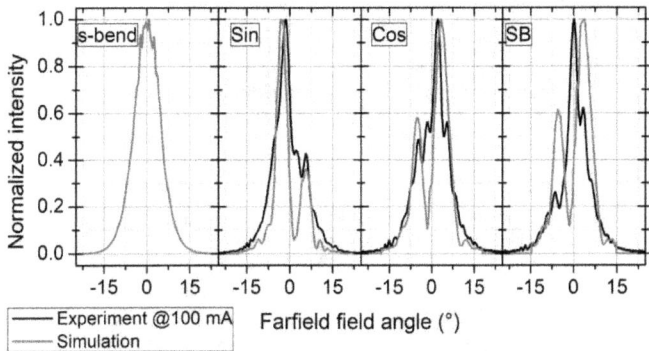

Figure 4.14: *Far field angles of the measured and of the simulated waveguide structures.*

The calculated transmission of the two main modes are given in Table 4.2. The straight RW has the highest transmission which is concentrated within the main mode, and does not allow transmission of other modes ($T < 0.1\%$). Similar values were obtained for the three S-bend structures, regardless of having a straight front section or not. This behaviour is expected as they all had similar spatial characteristics to the straight RW, see Fig. 4.9 and 4.10.

Table 4.2: *Simulated transmission coefficients for a straight, the three single S-bend and the three Y-branch RW structures (single arm).*

Structure	Transmission [%]	
	Mode 1	Mode 2
Straight RW	> 95	< 0.1
Sin S-bend	> 95	< 0.1
Cos S-bend	> 95	< 0.1
SB S-bend	> 92	< 0.1
Sin Y-branch	39	~11
Cos Y-branch	42	~10
SB Y-branch	38	~10

The three Y-branch RW structures however, have their transmitted power distributed within two modes. The Cos Y-branch has the highest transmission for mode 1 (main mode), followed by the Sin and then the SB, in agreement with previously published experimental results in [76]. Note that these values are calculated for the transmission through one of the two arms. In all three cases, the Y-branch structures allows the propagation of a second mode with up to $T \approx 10\%$.

The calculated transmission values can be related to the experimental behaviour observed in the output powers (see Table 4.1) and in the emission wavelength (see Fig. 4.4). The DBR-RW laser not only has a higher slope efficiency than the Y-branch lasers, but it also maintains single-mode operation (spectrally and spatially), over the entire investigated injection current range. In comparison, the Y-branch lasers had reduced output powers and beam quality, together with spectral side-modes at higher output powers. Thus suggesting that the transmission of a second mode can be the reason behind the reduced output powers and behind the multi-mode operation of the Y-branch lasers.

4.5 Device characteristics without operating the Y-section

The results so far indicate that the Y-section has a dominant impact on the performance of the Y-branch lasers. In this section, an alternative operation of these lasers is considered where the Y-section is not injected with current. As many of the upcoming investigations are similar to the ones already shown, this section will be relatively short.

4.5.1 Electro-optical characteristics

Previously, the three sections of the Y-branch lasers; Left- or right arm + Y + front section were all in a parallel connection. In this section, the same lasers are characterized but without connecting the Y-section. The contact scheme of the Y-branch laser can be seen in Fig. 4.2 and a photograph of the laser device can be found in Appendix E.

Fig. 4.15 shows the output power of a Sin based Y-branch laser at different injection currents through the left arm, obtained at different injection currents through the front section I_{front} and without operating the Y-section; $I_Y = 0$ mA. As can be seen, the threshold current is decreased and higher output powers are obtained with increasing I_{front}, as the front section becomes more and more amplifying. Note that similar performances were obtained for the Cos and SB Y-branch lasers when setting $I_Y = 0$ mA and when increasing I_{front}. The switching behaviour in Fig. 4.15 is explained by a saturation effect and a transition between a multimode LED state to a single lateral mode.

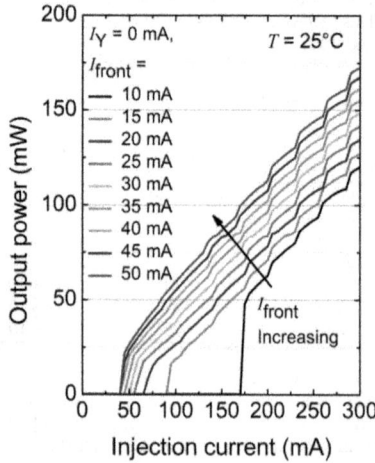

Figure 4.15: *Output power as function of the injection current through the left arm of a Sin Y-branch laser, with $I_Y = 0$ mA and at different front section currents.*

Based on the results of Fig. 4.15, this section consider the operation where $I_{front} = 50$ mA and $I_Y = 0$ mA, chosen due to the obtained output power. The output powers of the three Y-branch lasers as function of the injection current through each arm are shown in Fig. 4.16. Laser emission (reduced threshold) takes place at $I_{inj} = 15$ mA when injecting $I_{front} = 50$ mA. Maximum output powers of about $P_{max} = [229/265/243]$ mW at $I_{inj} = 500$ mA were obtained for the sin, cos and SB lasers, respectively. The slope efficiencies $S = [0.62/0.69/0.65]$ W/A were determined from linear fits in the range $[50, 300]$ mA. Overall, similar output power values were obtained when operating all three section of the Y-branch lasers, or as in this case where $I_Y = 0$ mA and $I_{front} = 50$ mA.

The corresponding emission wavelengths of the three Y-branch lasers are shown in Fig. 4.17, where single mode operation was maintained throughout the injection current range. This is a clear improvement to

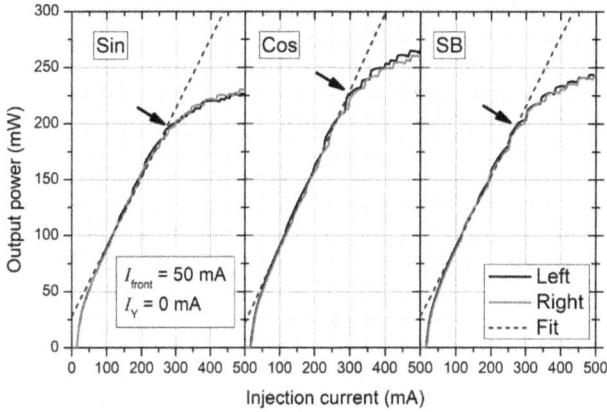

Figure 4.16: *Power characteristics of the investigated Y-branch lasers including linear fits, as function of the injection current through each laser arm at $T = 25°C$.*

the measurements in Fig. 4.4 where single mode operation was only maintained up to $I_{inj} < 400$ mA. This suggest that an unpumped Y-section provides a spectral filtering effect.

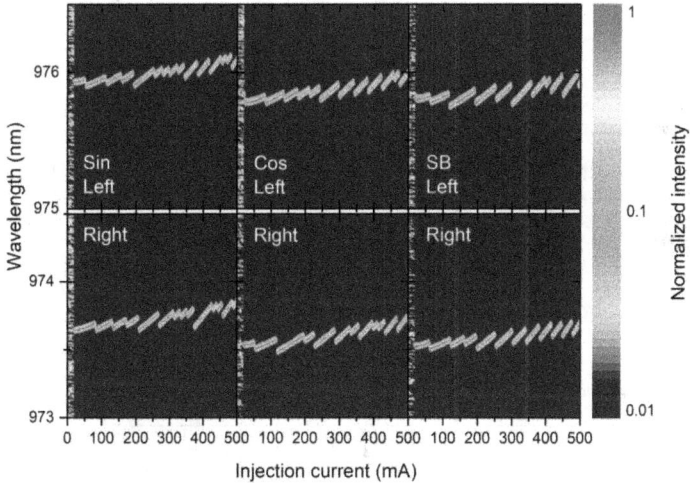

Figure 4.17: *False color contour plots of the emission wavelength of the studied lasers, as function of the injection current through each arm, with $I_Y = 0$ mA, $I_{front} = 50$ mA at $T = 25°C$.*

4.5.2 Spatial characteristics

The spatial characteristics are considered next, and the near field profiles of the Y-branch lasers are shown in Fig. 4.18 at different injection currents. Once again, side-modes (tails) are observed at low powers which gets suppressed at higher powers.

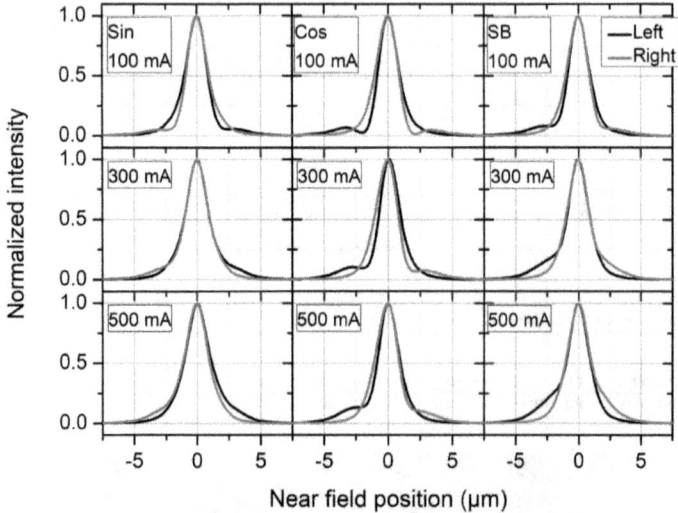

Figure 4.18: *Normalized near field profiles of the investigated lasers, at different injection currents through each arm, with $I_Y = 0$ mA, $I_{front} = 50$ mA at $T = 25°C$.*

The measured far field profiles are shown in Fig. 4.19 at different injection currents. Once again, the Sin laser has the majority of its intensity concentrated within the main peak, while the Cos and SB lasers show two intensity peaks. Whether the Y-section is operated or not, the far field performance is not influenced, see Fig. 4.7.

The obtained spatial characteristics are summarized in Table 4.3, where the near field widths seems more stable at increasing output powers when compared to the previous measurements in Table 4.1. The far field angles reduce and the M^2 values improves at increasing output powers, similarly to the results of Table 4.1.

Table 4.3: *Spatial characteristics of Y-branch lasers at different injection currents through each arm measured at $T = 25°C$, with $I_Y = 0$ mA and $I_{front} = 50$ mA.*

Laser at I_{inj} [mA]	Near field width [μm]	Far field angle [°]	Beam quality factor $M^2_{2.Mom.}$
Sin left: 100 / 300 / 500	6.1 / 6.3 / 6.3	21.3 / 19.9 / 19.5	1.8 / 1.8 / 1.7
Sin right: 100 / 300 / 500	6.1 / 6.1 / 6.1	23.1 / 20.4 / 19.9	2.0 / 1.8 / 1.7
Cos left: 100 / 300 / 500	6.9 / 6.9 / 6.9	24.3 / 22.5 / 21.7	2.4 / 2.2 / 2.1
Cos right: 100 / 300 / 500	6.6 / 6.3 / 6.3	23.3 / 22.0 / 19.9	2.2 / 2.0 / 1.8
SB left: 100 / 300 / 500	6.6 / 6.7 / 6.6	22.0 / 20.1 / 19.8	2.1 / 1.9 / 1.9
SB right: 100 / 300 / 500	6.2 / 6.3 / 6.2	21.3 / 18.5 / 17.9	1.9 / 1.7 / 1.6

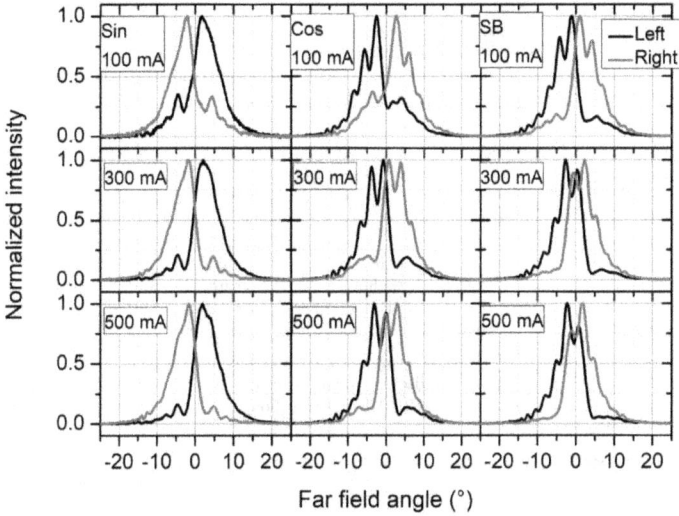

Figure 4.19: *Normalized far field profiles of the investigated lasers, at different injection currents through each arm, with $I_Y = 0$ mA, $I_{front} = 50$ mA at $T = 25°C$.*

4.5.3 Summary

The results of this section indicate that an unpumped Y-section has an advantage, acting as a spectral filter which provides single mode operation.

This being said, similar output powers and spatial characteristics were obtained under this operation setting of the Y-branch lasers. Although the Y-section was not operated, the generated light still had to pass through the Y-section, and is thus spatially affected by this section. The far field measurements suggest once again the Sin Y-branch laser is the one which resembles a Gaussian beam the most, with least changes at increasing output powers.

4.6 Conclusion

The experimental investigations of the three S-bend based Y-branch DBR-RW lasers show overall comparable performance between each other. A small difference is seen in the power characteristics, with the Cos based Y-branch DBR-RW laser providing about 6% more output power than the two other Y-branch lasers. Although this difference is small, it is in agreement with the simulated transmission coefficients obtained for passive RW structures.

On the other hand, when considering the spatial characteristics, the Sin based Y-branch DBR-RW laser is the favourite between the three lasers. This is seen by its far field profile as it resembles a Gaussian profile the most, and shows least changes with increasing output powers. The Sin based laser also have the overall better beam quality in form of the $M^2_{2.\mathrm{Mom.}}$ values. Therefore, for applications with demanding spatial characteristics, and where the far field profile is of great importance (e.g. optical imaging), the Sin Y-branch structure would be the laser of choice. This being said, the near field power distributions of all three Y-branch lasers are quite similar without a clear favourite. This suggests that in applications where the light is tightly coupled into a single mode fiber or in an optical amplifier (see chapter 8), all three Y-branch lasers should be equally suitable.

The overall performance of the SB based Y-branch laser lies between the Cos and Sin, having similar output power performance to the Sin, and comparable beam quality to the Cos. Interestingly, the SB have a narrower far field profile than the DBR-RW laser, although they have a similar ridge width. This might be interesting in application with specific or narrow far field requirements.

Throughout this investigation, it is clear that all three Y-branch lasers exhibit worse spatial performance when compared to both reference lasers, which is the "cost" of combining two S-bend sections. Passive RW simulations indicate that this effect is less dependent on the S-bend shape nor the front section, but rather the Y-section. At this point, the ridge becomes twice as wide, and this sudden "break" allows higher spatial modes to propagate through the waveguide. The result of this is seen in the spatial characteristics, but also in the electro-optical characteristics where a reduced slope efficiency and spectral multi-mode operation take place.

The impact of the Y-section on the spectral characteristics of the Y-branch laser was also investigated. Here, it was shown that the spectral characteristics can be significantly improved by simply not operating this section. This indicates that this section is responsible for the spectral multi-mode operation observed earlier, and a spectral filtering is obtained by not operating the Y-section; I.e. making it absorbent. This being said, the alternative operation setting ($I_Y = 0$ mA) did not improve the spatial characteristics significantly, as the light still had to pass through this section.

Finally, the simulated spatial characteristics can provide a qualitative description of the near fields, however they are not sufficient to fully describe the far field profiles. In addition, the spectral improvements obtained when setting $I_Y = 0$ mA indicates a connection between the Y-section and the spectral characteristics. Therefore, future work should include active cavity simulations of the Y-branch structures to not only understand the far field behaviour, but also the spectral characteristics.

Chapter 5

Multi-wavelength DBR-RW lasers

The main findings of the previous chapter indicate that the Y- (common) section strongly affects the overall laser performance. In this chapter, two other multi-wavelength lasers are investigated, namely four- and six-arm DBR-RW lasers. The motivation of this chapter is to investigate different common section structures and to study the influence of the bending curvature of the individual arms, while increasing the number of laser arms.

5.1   Four-arm DBR-RW Lasers

The waveguide scheme of the processed four-arm DBR-RW lasers is shown in Fig. 5.1. Each waveguide has its own 350 μm long DBR grating, and the four waveguides are joint (common section) before being emitted from the front section.

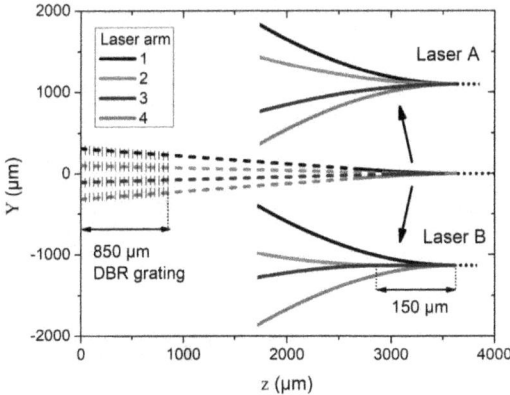

Figure 5.1: *Sketch of the investigated four-arm lasers. The close-ups shows the common sections of laser A and B.*

Each arm follows the SB structure consisting of an angled part, a single bend (SB) and a front section. The bend sections follow the Cos S-bend formula between $z = [L/2, L]$, see eq. (4.2). The half lateral offsets of the inner arms were $w_{\mathrm{inner}} = \pm 20$ μm for a bending length of $L_{\mathrm{inner}} = 1000$ μm. The outer arms were likewise defined between $z = [L/2, L]$, with half offset values of $w_{\mathrm{outter}} = \pm 60$ μm for a bending length of $L_{\mathrm{outter}} = 2000$ μm. The bend losses were once again reduced by implementing a large bend radius according to the etch depth and effective index step [77]. The inner arms have a bending radius of about 10.1 mm, while the outer have a bending radius of 13.5 mm.

The inner arms of the investigated lasers were angled by ±1.8° (similar to the SB Y-branch laser), while the outer were angled by ±3.4°. These values provide a smooth transition between the angled and the bend sections, see Fig. 5.1. The DBR gratings were equally angled.

The SB structure is used here to combine the four arms as it provides low curvature with low bending losses. In addition and more importantly, the angled RW sections ensures further displacement between the individual arms at the back side of the devices. This provides more space for the processing of the individual DBR gratings and the micro-heaters as will be shown in chapter 6. The motivation of these lasers is to compare two different combiner section and investigate the influence of having two different curvatures of the inner and outer arms.

In this investigation, two four-arm lasers (laser A and B) are investigated with different common sections. Laser A has a single intersection point for all four waveguides, while laser B has two separate points. In the latter case, the inner and outer arms' intersection points were displaced by 150 μm from one another, see Fig. 5.1. This displacement is compensated by shortening the angled sections. At position $z = 0$ μm in Fig. 5.1, the outer arms of laser A are positioned at $Y = \pm295.6$ μm while the inner were at $Y = \pm102.2$ μm, respectively. In the case of laser B, the inner arms positioned at $Y = \pm309.8$ μm while the outer are at $Y = \pm102.2$ μm.

The contact scheme of these lasers is shown in Fig. 5.2. The individual sections (each arm, common- and the front section) have isolated gold surface contacts to control the current injections separately. A photograph of the laser can be seen in Appendix E.

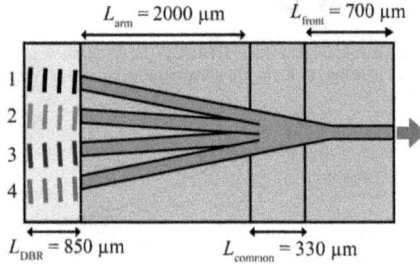

Figure 5.2: *Contact scheme of the investigated four-arm DBR-RW lasers.*

The individual arms of laser A are denoted as; A1, A2, A3 and A4 and likewise for Laser B; B1, B2, B3 and B4, see Fig. 5.2. Note that the inner and outer arms of both lasers have different curvatures (as they have different w and L values) and the effect of this will also be considered in the following.

5.1.1 Electro-optical characteristics

The PUI curves of the two lasers are shown side-by-side in Fig. 5.3 when operated at a heat sink temperature of $T = 25$°C. Similarly to the Y-branch lasers (see Sec. 4.5), the common section was not operated in this investigation in order to obtain a complete single mode operation range. In addition, the front section was injected with $I_{front} = 35$ mA. The injection current through each arm was varied between $I_{inj} = [0, 500]$ mA with a step size of $\Delta I_{inj} = 5$ mA.

The four arms of each laser were characterized and the threshold current I_{th}, the maximum measured output power P_{max} and the slope efficiency S were obtained, see Table 5.1. The slope efficiencies were obtained through linear fits between $I_{inj} = [100, 300]$ mA, a region above threshold and below the role-over effect. Table 5.1 indicates a clear difference in performance when comparing the outer (A1 & A4) and the inner arms (A2 & A3) of laser A. This is seen in I_{th}, P_{max} and S, which is also clear in Fig. 5.3(a). Overall, a worse performance is observed for the outer than for the inner arms.

The laser characteristics of the four arms of laser B were likewise obtained and are summarized in Table 5.1. In this case, similar performance was observed with a smaller discrepancy between the outer

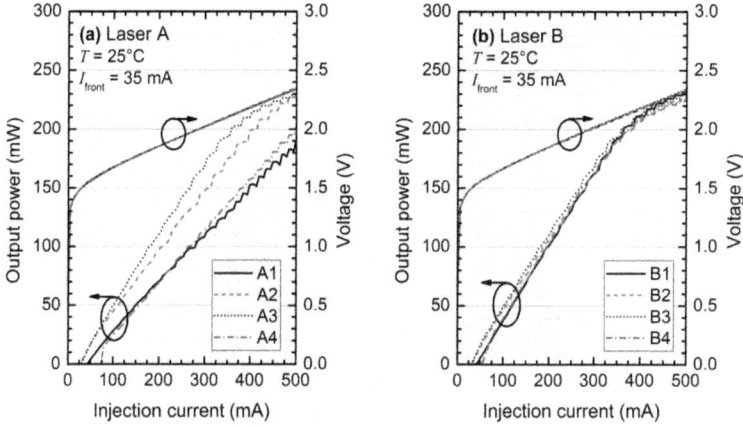

Figure 5.3: *PUI characteristics of the individual arms of Laser A and of Laser B, respectively.*

arms (B1 & B4) and the inner arms (B2 & B3), which is also clear in the power curves in Fig. 5.3(b). This indicates that the joining section has a strong impact on the output power, while the curvature plays a minor role as seen for laser B.

Table 5.1: *Laser characteristics of the two four-wavelength DBR-RW lasers.*

Laser at $T = 25°C$	I_{th} [mA]	S [W/A]	P_{max}(500 mA) [mW]	$\lambda(P_{max})$ [nm]
A1	50	0.41	190	978.01
A2	30	0.52	225	975.74
A3	35	0.60	230	973.63
A4	(75)	0.45	200	971.28
B1	45	0.63	230	978.14
B2	35	0.60	225	975.87
B3	35	0.60	230	973.63
B4	(60)	0.63	225	971.44

An irregular power curve is seen for A4 and B4 which is why their I_{th} values are written in parenthesis in Table 5.1. By back-interpolation of these power curves, expected threshold values of about 46 mA and 39 mA were obtained for A4 and B4, which are closer to the threshold values of A1 and B1 (the corresponding pairs). This switching behaviour is likely explained by a transition between a multimode LED state to a single lateral mode.

Next, the spectral behaviour of the two lasers is investigated, see Fig. 5.4. The two lasers emits at wavelengths of about $\lambda_1 = 978$ nm, $\lambda_2 = 975$ nm, $\lambda_3 = 973$ nm and $\lambda_4 = 971$ nm, respectively. As can be seen in Fig. 5.4, all eight arms of the two lasers maintain single-mode operation over the investigated injection current range. This was obtained because the common section was not operated; $I_{commom} = 0$ mA. The discrepancy in the threshold currents between the inner and outer arms of laser A (see Fig. 5.3(a)) can also be observed in the spectral emission. Likewise, the observed threshold values are in agreement with the power curves, see Fig. 5.3(b).

The individual optical spectra of the investigated lasers at P_{max} are shown in Fig. 5.5, and the peak emission wavelength values are summarized in Table 5.1. From this measurement, the spectral linewidth of all eight measurements was estimated to be < 17 pm, limited by the resolution of the spectrometer. This indicates that the common section and the different curvatures of the inner/outer arms have within the studied parameter ranges no impact on the spectral performance.

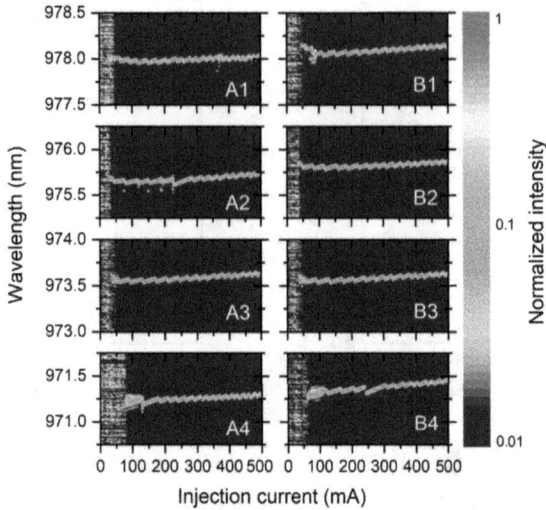

Figure 5.4: *False colour contour plots of the emission wavelength of the investigated lasers, as function of the injection current with $I_{\text{common}} = 0$ mA, $I_{\text{front}} = 35$ mA at $T = 25°C$.*

The spectral distance between the four arms was processed to be $\bar{2}.2$ nm. In addition, the DBR gratings were designed to provide similar wavelengths from both lasers (A1 and B1, A2 and B2, ect.), and thus the observed small mismatch between the wavelengths in Fig. 5.5 are due to manufacturing tolerances.

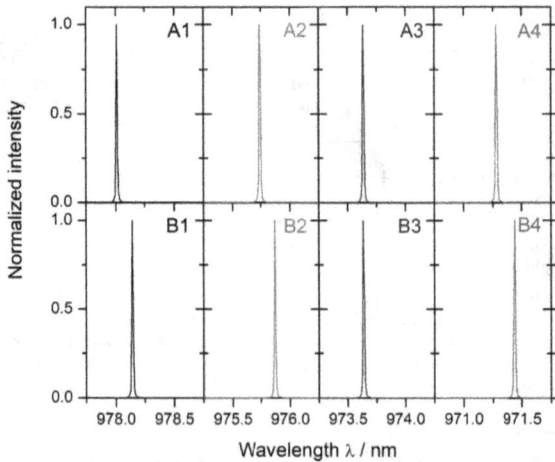

Figure 5.5: *Normalized individual optical spectra of the investigated lasers at P_{max}, measured at $I_{\text{inj}} = 500$ mA, $I_{\text{common}} = 0$ mA, $I_{\text{front}} = 35$ mA at $T = 25°C$.*

The output power behaviour in Fig. 5.3 suggest different performances from the inner and outer arms in the case of laser A, and comparable performance in the case of the four arms of laser B. This is explained by the different common section designs. The comparable values for the arms of laser B indicate that the curvature of the individual arms has a minor effect on the electro-optical performance.

5.1.2 Spatial characteristics

The spatial characteristics of both four-arm lasers were obtained according to the moving slit method, and the normalized lateral intensity distributions of the near fields (at $I_{inj} = 200$ mA) are shown side-by-side in Fig. 5.6.

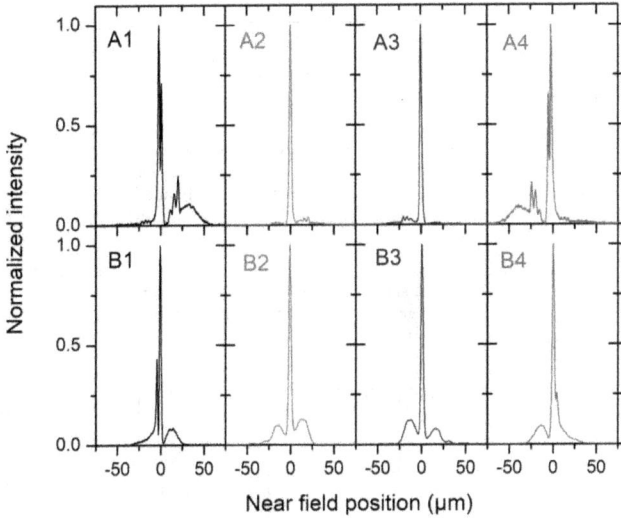

Figure 5.6: *Normalized near field profiles of the investigated lasers at $I_{inj} = 200$ mA, $I_{common} = 0$ mA, $I_{front} = 35$ mA at $T = 25°C$.*

The near fields of the inner arms of laser A (A2 and A3) show defined central lobes, with a number of low intensity side-lobes. In the case of A2, these side-lobes are mainly concentrated on the right-hand-side (positive position), and vice versa for A3. The outer arms of laser A (A1 and A4) shows similar behaviour, having defined central lobes but with strong side-lobes concentrated toward the inner arms.

The inner arms of laser B show two broad neighbouring peaks and defined central lobes in between. Similarly to the inner arms of laser A, these neighbouring peaks are stronger towards the aperture side. The outer arms of laser B show similar behaviour to those of laser A, however with weaker side-lobes that are also concentrated toward the inner arms.

The near fields of both the inner and outer arms of both lasers have mirror symmetry, and show the effect of having stronger side-lobes towards the aperture (front section).

The near field widths of both lasers are summarized in Table 5.2. In the case of laser A, comparable values were obtained for the inner arms (~ 5 μm) and for the outer arms (~ 26 μm). In the case of laser B, the same behaviour was observed, however with a smaller discrepancy between the inner (~ 6 μm) and outer arms (~ 9 μm).

By considering the near field distributions, the power in central lobe P_{CL} of each arm was obtained, see Table 5.2. Once again, laser A have comparable values for the inner (or outer) arms, but with a large discrepancy from 32 to 78% between those. In the case of laser B, the P_{CL} values of the inner and outer arms only vary between 45 to 53%. Note that this parameter was not considered previously, as

Table 5.2: *Spatial characteristics of the investigated four-arm lasers, at $I_{\text{front}} = 35$ mA, $I_{\text{common}} = 0$ mA and $T = 25°C$.*

Laser at $I_{\text{inj}} = 200$ mA	Near field width [μm]	Far field angle [°]	M^2_{1/e^2}	P_{CL} [%]
A1	26	22	8.2	32
A2	5	22	1.5	78
A3	5	19	1.4	78
A4	27	22	8.4	36
B1	7	21	2.2	45
B2	6	21	1.6	53
B3	6	20	1.7	51
B4	9	22	2.8	53

the reference and the Y-branch lasers had most of their power concentrated within the central lobe, with values above $P_{\text{CL}} > 90\%$.

The results in Fig. 5.6 and in Table 5.2 indicate that in the case of laser B, comparable near fields were obtained with matching widths. In the case of laser A however, it seems that the outer arms have a spatial filtering effect on the inner arms as they have much smaller side-lobes in comparison to the inner arms of laser B. In addition, the inner arms of laser A acts oppositely and have a degrading effect on the outer arms. This is seen as the outer arms of laser A have stronger and wider side-lobes in comparison to the outer arms of laser B.

The measured far field profiles of the two lasers are shown in Fig. 5.7. The far field profiles of both lasers consist of multiple high intensity peaks, showing mirror symmetry between the inner and outer arms. By using the far field angles of the investigated lasers together with the near field values, the propagation values M^2_{1/e^2} in Table 5.2 were obtained.

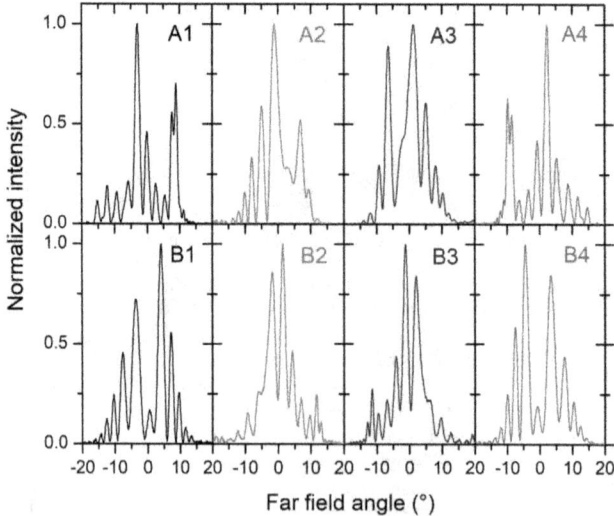

Figure 5.7: *Normalized far field profiles of the investigated lasers at $I_{\text{inj}} = 200$ mA, $I_{\text{common}} = 0$ mA, $I_{\text{front}} = 35$ mA at $T = 25°C$.*

In the case of laser A, the inner arms have comparable M^2_{1/e^2} values (~ 1.5) and likewise for the outer arms (~ 8.2). The same effect was observed in the case of laser B, however with a smaller discrepancy in the M^2_{1/e^2} values, ranging between 1.6 for the inner and 2.8 for the outer arms, respectively.

Note that at the chosen injection current of $I_{inj} = 200$ mA, an output power of about $P = 100$ mW is emitted, which is the typical value needed in different applications. At higher injection currents, the beam quality deteriorates for both lasers, however the overall observed behaviour is maintained, i.e. the large discrepancy between the inner and outer arms for laser A and the more comparable behaviour observed for laser B.

5.1.3 Summary

In this study, two different four-arm DBR-RW lasers were investigated; Laser A having a single inter-section and laser B with two separate intersection points between the individual arms. Through this investigation, it is clear the two structures exhibit different laser characteristics. This was observed in the output power characteristics, where laser A have distinct performance when comparing the inner and outer arms. In particular, the outer arms had smaller slope efficiencies and lower output powers in comparison to the inner arms. On the other hand, laser B showed similar slope efficiencies and power values for all four arms.

Similar behaviour was observed in the spatial characteristics, where Laser A had a higher M^2_{1/e^2} values (~ 8.2) for the outer arms than for the inner arms (~ 1.5), and likewise smaller P_{CL} (32%) for the outer arms when compared to the inner (78%). The four arms of laser B had more comparable beam qualities, where the M^2_{1/e^2} values only ranged between 1.6 to 2.8 and the P_{CL} ranged between 45 to 53%.

The output power and spatial characteristics indicate that the main difference is due to the intersection point, and is less influenced by the different curvature of the inner and outer arms. This is clear from the performance of laser B, while the discrepancy seen between the four arms of laser A (similar curvature to laser B) is caused by the single intersection point.

On the other hand, the spectral behaviour was not influenced by the laser structures, at least within the studied parameters and spectral resolution. Single mode operation and a spectral width smaller than 17 pm (resolution limit) were maintained over the investigated injection current range for both lasers.

The spatial characteristics of both lasers, especially the far field profiles, indicates non-diffraction limited beams which is the "cost" of having four monolithically combined laser arms. This is believed to be caused by the joining sections, where in the case of laser A the RW suddenly becomes 4 times as wide, and in the case of laser B becomes twice as wide at the first joining point, and 3 times as wide at the second joining point. This is believed to be the reason behind the difference in the overall performance when comparing the two lasers. In both lasers, this change allows higher order (spatial) modes to propagate as seen in the near and far field measurements. Nonetheless, in the targeted applications where a spatial filtering takes place, the observed spatial characteristics and output powers makes both lasers suitable in MOPA systems.

One interesting result from laser A is that the single intersection provided a spatial filtering effect on the inner arms' near fields, and a degrading effect on the outer arms, leading to the different spatial characteristics. This spatial filtering effect could be interesting to study further, as it could potentially lead to improved spatial performance of multi-arm lasers.

Finally, it is clear that laser B is preferred over laser A in terms of homogeneous output power and spatial characteristics between each arm. In this section, no simulations were presented as the numerical model used in chapter 4, did not provide meaningful results when used to simulate passive four-arm structures. Future work should therefore include active simulations of the intersection between the RW in four-arm lasers. This can potentially bring further understanding of the experimental results shown in this study, and bring further improvement to the spatial characteristics of multi-arm DBR-RW lasers.

5.2 Six-branch DBR-RW laser

The findings of the previous section indicate that the laser structure with individual intersection points is preferred over the one with a single point. In addition, the different curvatures of the inner and outer arms played a minor role on the overall laser performance. In this section, another multi-wavelength laser namely a six-arm DBR-RW laser will be considered, with three separate intersection points. The contact scheme of the six-arm laser is seen in Fig. 5.8(a), where each arm has its own DBR grating. The six arms are joint before being emitted from a single aperture (front section). In this structure, no common section was implemented mainly due to space constriction. A photograph of this laser can be found in Appendix E showing how the injection current was lead into the different arms and front section. A zoom of the the three separate intersection points is shown in Fig. 5.8(b), which are spaced 100 μm apart.

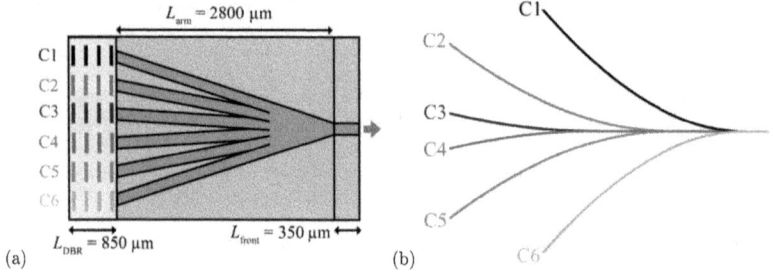

Figure 5.8: *Sketch of (a) the six-arms DBR-RW laser and (b) close-ups of the intersection points.*

The arms follow the SB structure once again, consisting of an angled part, a SB and a front section. The bend sections follow the Cosine S-bend eq. (4.2), with half lateral offset values of $w_{outer} = 200$ μm (C1 and C6), $w_{middle} = 120$ μm (C2 and C5) and $w_{inner} = 40$ μm (C3 and C4), for bending lengths of $L_{outer} = 2000$ μm, $L_{middle} = 2000$ μm and $L_{inner} = 2000$ μm, respectively. The bend structures are defined between $z = [L/2, L]$, and the lateral offsets at the back of the device are: $\pm[532, 310, 98]$ μm for the outer, middle and inner arms respectively. The outer arms were angled by $\pm 8.9°$, the middle by $\pm 5.4°$ and the inner were angled by $\pm 1.8°$, respectively, and the corresponding DBR gratings were angled equally. These values were found to provide a smooth transition between the angled and the bend sections. The bending radii are $R_{inner} = 28.7$ mm, $R_{middle} = 9.6$ mm and $R_{outer} = 5.8$ mm, respectively.

As described earlier, the motivation behind using the SB structure is the smooth combination between the individual arms, while the angled sections provide an increased spacing between the individual DBR gratings for easier manufacturing and heater implementation.

5.2.1 Electro-optical characteristics

The PUI characteristics of the individual arms of the Six-arm laser were obtained in a similar fashion to the previous lasers. This was done for injection currents between $I_{inj} = [0, 500]$ mA for $\Delta I_{inj} = 5$ mA, with $I_{front} = 35$ mA at a heat sink temperature of $T = 25°C$, see Fig. 5.9(a).

The threshold current I_{th}, maximum measured output power P_{max} and the slope efficiency S, fitted between $I = [100, 300]$ mA, were obtained and are summarized in Table 5.3. As can be seen, comparable threshold current values were obtained from the individual arms. In the case of the slope efficiencies and the maximum output powers, larger deviations were observed with S values varying between 0.47 and 0.63 W/A and P_{max} between 199 and 246 mW. This deviation is observed for C2 and C5, which provide the highest and lowest output powers, see Fig. 5.9(a). The difference between C2 and C5 is also somewhat peculiar as this pair of arms are symmetrical with each other, and could indicate a defect/error during the manufacturing process or in the design layout.

The spectral behaviour was likewise obtained for the individual arms, see Fig. 5.9(b). Each of the six arms shows single- and multi-mode operation, where in the latter maximum spectral widths of about 262 pm at $1/e^2$ and around 77 pm at FWHM were measured.

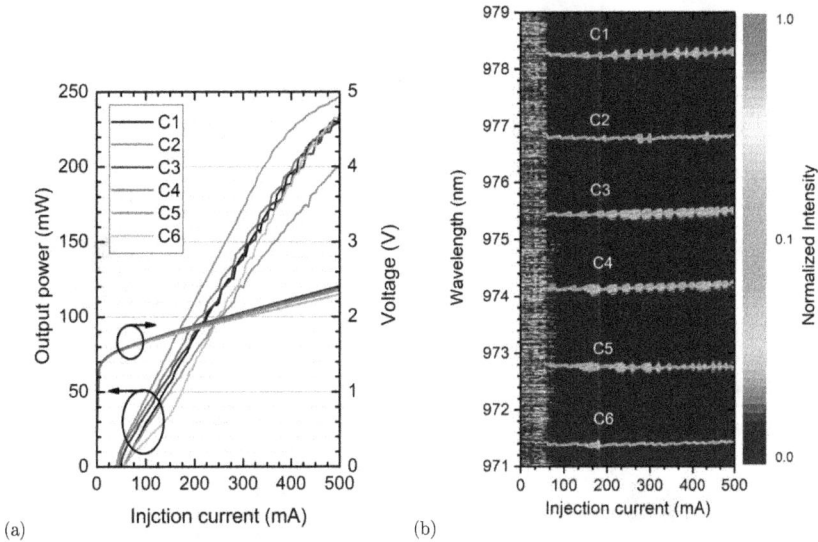

(a)   Injction current (mA)   (b)   Injection current (mA)

Figure 5.9: *PUI characteristics (a) and the corresponding emission wavelength (b) of the six-arms, as function of the injection current, measured at $I_{\text{front}} = 35$ mA at $T = 25°C$.*

Table 5.3: *Laser characteristics of the six-wavelength DBR-RW lasers.*

Laser at $T = 25°C$	I_{th} [mA]	S [W/A]	$P_{\text{max}}(500$ mA) [mW]	$\lambda(P_{\text{max}})$ [nm]
C1	55	0.56	234	978.52
C2	55	0.47	199	977.11
C3	45	0.52	228	975.78
C4	45	0.53	231	974.48
C5	40	0.63	246	973.02
C6	50	0.58	235	971.67

a spectral width (at FWHM) of up to 120 pm was observed. This multi-mode behaviour is likely explained by the lack of a separate common section contact, as was the case in both the Y-branch and in the four-arm lasers. Due to manufacturing space limitations, this section was not processed on the six-arm lasers.

The individual spectra at P_{max} are shown in Fig. 5.10, indicating single-mode operation at maximum output powers. The corresponding emission wavelengths are summarized in Table. 5.3.

5.2.2   Spatial characteristics

The spatial characteristics of the individual arms at $I_{\text{inj}} = 200$ mA were obtained in a similar fashion to the previous sections, and the near and far field distributions are shown in Fig. 5.11.

The near field profiles show well defined central lobes, with different neighbouring peaks that are largest for the outer arms and reduces for the inner arms. Similarly to the four arm lasers, a symmetry is observed between each pair of arms, and the neighbouring peaks for all six lasers are strongest toward the aperture side.

The near field widths are summarized in Table 5.4 which show comparable values for the four inner arms (~ 20 μm) and larger values for the outer two arms C1 and C6 of 30 and 40 μm, respectively. This tendency was also observed in the power in central lobe values P_{CL} as summarized in Table 5.4.

43

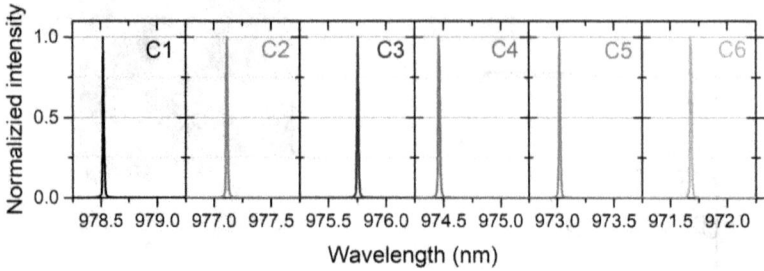

Figure 5.10: *Normalized optical spectra of the six-arms near P_{max} measured at $I_{front} = 35$ mA and $T = 25°C$.*

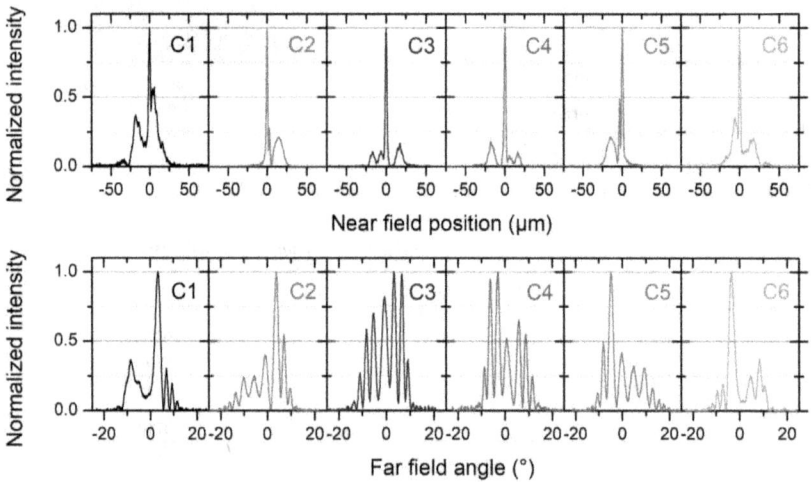

Figure 5.11: *Normalized near field positions (top) and far field angles (bottom) of the six-arm DBR-RW laser at $I_{inj} = 200$ mA, $I_{front} = 35$ mA and $T = 25°C$.*

Table 5.4: *Spatial characteristics (using the $1/e^2$ definition) of the six-wavelength DBR-RW lasers.*

Laser	Near field width	Far field angle	M_{1/e^2}^2	P_{CL}
at $I_{inj} = 200$ mA	[μm]	[°]		[%]
C1	32.0	20.8	9.4	26
C2	21.8	22.2	6.8	50
C3	21.4	21.5	6.5	51
C4	20.6	21.5	6.2	51
C5	21.9	21.2	6.5	42
C6	40.9	20.8	12.0	34

The far field profiles shown in Fig. 5.11 indicate mirror symmetry between each pair of arms. Multiple high intensity peaks are seen, similarly to the far field profiles of the four-arm lasers. The far field angles (obtained at $1/e^2$) are summarized in Table 5.4 together with the corresponding beam propagation factors M^2_{1/e^2}. Once again, comparable behaviour for the four inner arms are observed with M^2_{1/e^2} values ranging between 6.2 and 6.8, while the outer arms, C1 and C6, have M^2_{1/e^2} values of 9.4 and 12, respectively. This discrepancy is likely due to the stronger curvature of the outer arms. While this effect is clear in the beam quality comparison, it is less evident in the output powers of the individual arms, which is likely reduced by the amplifying front section.

Once again, the beam quality deteriorates at higher injection currents, however the overall observed behaviour is maintained e.g. the larger discrepancy for the outer two arms (C1 and C6) and the more comparable values for the four inner arms (C2, C3, C4 and C5). The injection current of $I_{\mathrm{inj}} = 200$ mA was chosen as it provides an output power of about $P = 100$ mW, which is sufficient in multiple applications.

5.2.3 Summary

In this section, a six-arm DBR-RW laser was investigated with three separate intersection points. The six arms have similar threshold current values and all provided output powers above 200 mW, with some smaller discrepancy in the electro-optical characteristics.

The six-arms showed broad regions of multi-mode emission, with single-mode operation at some injection current regions. This puts some restrictions on this device when choosing a certain wavelength and corresponding output powers. This is in contrast to the four-arm lasers which showed single mode operation at all output power. This difference is believed to be caused by the lack of an isolated common section in the six-arm device, which was implemented in the Y- and the four-arm lasers. In those cases, an absorbent($I_Y = 0$ mA) common section provided a spectral filtering effect. Such a section was not included in the six-arm laser.

Unsurprisingly, the spatial characteristics of the six-arms laser showed non-diffraction limited beams, which was also the case the four-arm lasers. This behaviour is the cost of having multiple monolithically combined laser arms.

The outer arms had higher M^2_{1/e^2} values (9.4 and 12.0) whereas the four inner arms had comparable values (between 6.2 and 6.8). This discrepancy is believed to be caused by stronger bending losses, due to the stronger curvature of the outer arms. This being said, this discrepancy is still smaller than in the case of the four-arm laser with a single intersection point (laser A). In addition, the M^2_{1/e^2} values of the four inner arms are comparable with one another, similarly to the case of the four-arm laser with two intersection points (Laser B).

Finally, although the six-arm laser had large M^2_{1/e^2} values and low P_{CL}, the well-defined central lobes in the near field profiles together with the relatively high output powers, makes this laser suitable for MOPA systems. In such systems, a spatial filtering takes place and usually require relatively small saturation powers.

5.3 Conclusion

In this chapter, two concepts of multi-arm DBR-RW lasers were investigated. The first section compares two four-arm lasers, one with a single intersection point between the four arms (Laser A), and another with two separate points (Laser B). Laser A showed different performances for the inner and outer arms, both in terms of output power and spatial characteristics. Laser B on the other hand showed comparable performance for the inner and outer arms. This indicates that the intersection point is responsible for the performance of these lasers, similarly to the case of the Y-branch lasers. In addition, both four-arm lasers have different curvature for the inner and outer arms. This had a minor effect on the laser characteristics, as comparable performances were obtained from all four arms of laser B.

To further investigate the effect of the RW curvature, a six-arm laser with three different curvatures and three individual intersection points was developed and investigated. This laser provided comparable output power values with some smaller discrepancy between the arms. Spatially, the outer arms with the largest curvatures had worse beam qualities ($M^2_{1/e^2} = 9.4$ and 12.0) when compared to the four inner arms ($M^2_{1/e^2} \sim 6.5$). This being said, this discrepancy is still smaller than the one observed in the case of the four-arm laser with a single intersection point (laser A). In comparison to the four-arm lasers, multiple regions of spectral multi-mode were observed from the six-arm laser. This is believed to be caused by the lack of an isolated common section, which in the case of the Y- and the four-arm lasers serve as a spectral filter.

The introduction of more laser arms reduces the beam quality as more and more spatial modes are supported, see Fig. 5.12. In some application which needs both wide tuning and excellent beam quality, the approach of multi-arm lasers introduces some challenges. However, in the targeted MOPA applications, these light sources could still be suitable. Although the beam quality deteriorates and the power in central lobe P_{CL} decreases with increasing number of arms, these light sources still provide enough powers to saturate a PA with well defined beam waits. Even in the case of $P_{CL} < 30\%$, having more than 200 mW should still provide enough power to properly seed and saturate a PA.

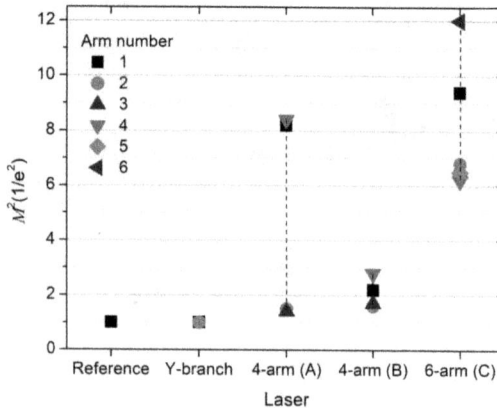

Figure 5.12: *Beam quality factors of the different lasers investigated so far.*

Finally, to fully understand and improve the performance of multi-arm lasers, active cavity simulations are required. Passive waveguide simulations could not fully describe the Y-branch laser performance, and even less in case of the four- and six-arm lasers. Although each laser arm is operated separately, it is reasonable to assume that interaction between the different arms takes place at the intersection point, which affects both the spectral and the spatial performances.

Chapter 6

Heater based wavelength tuning

The motivation behind the multi-wavelength lasers is to obtain an extended wavelength tuning compared to a "regular" DBR-RW laser. The tuning of these devices can be either obtained electrically or thermally.

Electrical tuning provides a shift in wavelength (decreasing) by inducing carriers into a passive grating section, which reduces the effective index of refraction (see Sec. 2.6). This tuning mechanism is employed in InP based semiconductor lasers (at 1.3 or 1.55 μm) [41, 81]. This effect however remains a challenge in GaAs based semiconductors (700 − 1100 nm), and only sub-nanometer of electrical tuning has been demonstrated [64].

Thermal tuning on the other hand, introduces a temperature gradient which slightly increases the index of refraction, resulting in a wavelength shift (increasing) at higher temperatures. This mechanism is typically obtained by resistor based micro-heaters, embedded on top of the grating section [82, 83].

Electrically controlled micro-heaters will be considered in this chapter, together with the material heating properties, heater configurations and tuning reproducibility. This is followed by the tuning characterization of the individual lasers; the DBR-RW reference laser, the three types of Y-branch lasers (Sin, Cos and SB), the two four arms laser (Laser A and B) and finally the six-arm laser (Laser C).

6.1 Heater configurations

Before investigating the tuning properties of the micro-heaters, the tuning coefficient of the developed process is estimated. This is done by measuring the emission wavelengths at different (heat sink) temperatures. Fig. 6.1 shows the emission wavelengths of a Y-branch DBR-RW laser operated at different injection currents at different heat sink temperatures. Due to the linear relationship between the wavelength and temperature, a slope of $\Delta\lambda/\Delta T = (65.0 \pm 0.5)$ pm/K is estimated from linear fits of the different measurements.

Next, the micro-heaters are considered, in particular the two embedded on top of each of the two DBR-gratings in the Y-branch lasers, see Fig. 6.2. A left and a right heater are implemented, each with its own two contact pads. This investigation consider the operation of the left arm of a Y-branch laser, but with four different heater configurations: Left heater, right heater, both heaters in a parallel connection and in series. Fig. 6.3(a) shows the emission wavelength as function of the heater current for these four cases. As can be seen, the laser emits around 974.73 nm and different tuning ranges were obtained, as summarized in Table 6.1.

When operating the left heater, a tuning range of 2.57 nm is obtained at a heater current of 600 mA. When operating the right heater, a tuning range of only 0.90 nm is measured, due to the (physical) distance between the left DBR grating and the right heater. When both heaters are in parallel connection, an even smaller tuning range of 0.73 nm is obtained. This is due to the reduced resistance;

$$R_{\text{parallel}} = \frac{1}{R_{\text{left}}} + \frac{1}{R_{\text{right}}} \ , \tag{6.1}$$

47

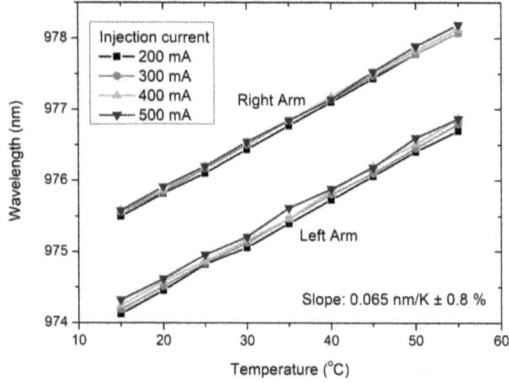

Figure 6.1: *Emission wavelength of a Y-branch DBR-RW at different injection currents as function of heat sink temperature.*

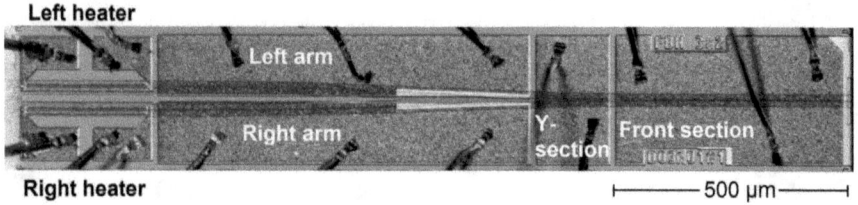

Figure 6.2: *Photograph of a Y-branch DBR laser with the corresponding bonding scheme of the individual sections. The two micro-heaters are seen on the left, each consisting of two contact pads.*

obtained in a parallel connection. The widest tuning of 3.90 nm is obtained when the two heaters are connected in series, due to the highest obtained resistance of

$$R_{\text{serie}} = R_{\text{left}} + R_{\text{right}} \ . \tag{6.2}$$

The super imposed emission spectra in Fig. 6.3(a) indicate single mode operation during the wavelength tuning in all four heater configurations. A quadratic tuning behaviour is seen, as the induced heater power scales quadratically with the heater current; $P = I^2 R$. The induced temperature gradient ΔT is obtained by dividing the tuning range by the material tuning coefficient of 65 pm/K, see Table 6.1.

Table 6.1: *Tuning characteristics of the left arm at different heater configurations.*

Connection	I_{heater} [mA]	U_{heater} [V]	P_{heater} [W]	$\Delta\lambda$ [nm]	ΔT [K]	ΔP [%]
left heater	600	1.68	1.0	2.57	39	2.38
right heater	600	1.72	1.0	0.90	13	0.83
Both heaters - parallel	600	0.72	0.4	0.73	11	0.62
Both heaters - series	600	3.57	2.1	3.90	60	4.61

Fig. 6.3(b) shows the output power and heater voltage during the wavelength tuning. A small power variation is seen when operating the right heater and under parallel connection. This is an agreement with the narrow tuning range obtained from these two settings. In the case of the left heater, a stronger

48

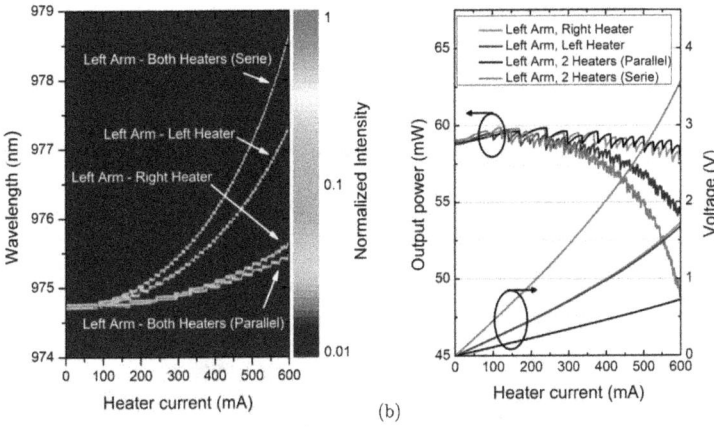

(a)   Heater current (mA)   (b)   Heater current (mA)

Figure 6.3: *Tuning characteristics showing (a) the emission wavelength and (b) the output power and heater voltage at different heater configurations as function of the heater current at $T = 25°C$.*

power drop of about 2.4% is observed during the wavelength tuning, and an even stronger drop of 4.6% is obtained for the "in series" connection. This is once again in agreement with the corresponding tuning ranges of the individual heater settings.

The heater voltage drop in Fig. 6.3(b) indicates once again that in the case of wide wavelength tuning, a higher voltage is measured which corresponds to a higher induced heater power: $P = I\, U$. These power values are summarized in Table 6.1. It is seen that the left and right heaters have similar voltage drops, indicating comparable heating performances. In summary, the higher induced heater power, the higher the temperature gradient, the wider the wavelength tuning and the stronger power drop during the tuning.

Finally, Fig. 6.4 shows the individual emission spectra during the wavelength tuning using the in series heater connection. Single mode operation is once again confirmed, with emission widths smaller than 17 pm (resolution limited) are maintained.

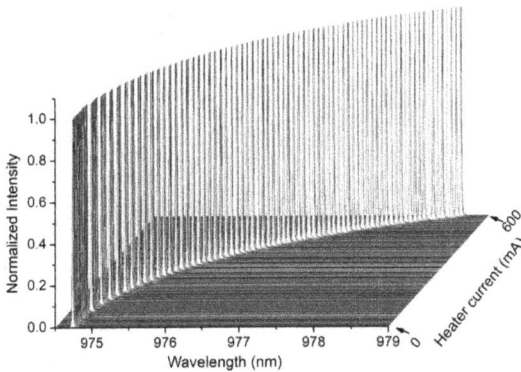

Figure 6.4: *Individual emission spectra obtained during the wavelength tuning at $T = 25°C$.*

49

6.2 Tuning reproducibility

Reproducibility is an important parameter when considering wavelength tuning. This is considered in this section by measuring the output power and emission wavelength during the wavelength tuning at different times. Fig. 6.5(a) and (b) show the output power and heater voltage drop of the left and right arms of a Y-branch laser respectively, measured at different times. A small variation is seen in the output power, which is shown as the standard deviation of mean (SDOM) of the emitted output power in Fig. 6.6(a). A maximum SDOM of about 0.3 mW for the left and 0.4 mW is observed for the right arm, during the 6 and 7 measurements, respectively. These measurements were obtained using a Gentech *PH100-Si UV* power meter with a noise equivalent power of 10 pW [84]. The measurements of the left arm were carried over two consecutive days, which explains the higher SDOM in the output power in comparison to the right arm. This is also seen in Fig. 6.5(a) where the individual measurements do not overlap as good as in the case of the right arm.

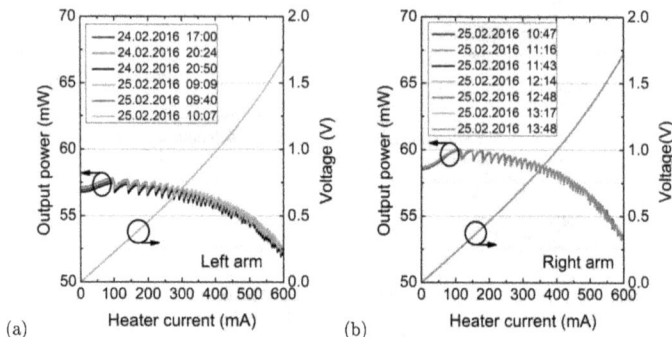

Figure 6.5: *Output power and heater voltage characteristics of (a) the left arm and (b) the right arm of a Y-branch laser as function of the heater current, obtained at different times at $T = 25°C$.*

Likewise, the emission wavelength during the tuning is measured over multiple measurements, and Fig. 6.6(b) shows the SDOM of the emission wavelengths of the left and right arms, respectively. As seen, a maximum variation of about 25 pm is seen in the emission wavelength during the different measurements. Note that the mode-spacing of this laser is 52 pm (see Sec. 4.1), thus indicating a high level of reproducibility which is about twice the 17 pm resolution limit. The individual peaks in Fig. 6.6(b) are likely explained by slightly shifted mode hops that are averaged over the different measurements.

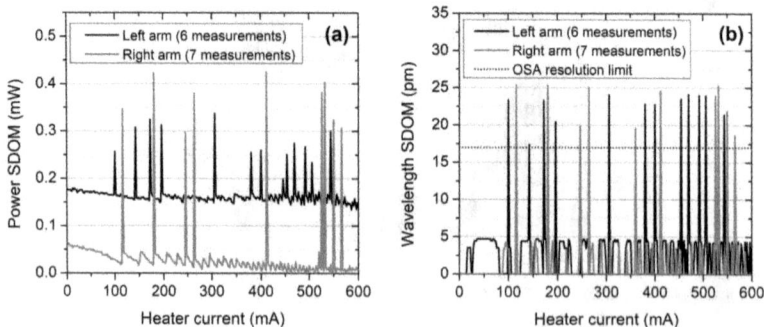

Figure 6.6: *SDOM of (a) the output power and (b) the emitted wavelength of the left and right arms, respectively, as function of the heater current obtained over 6-7 individual measurements at $T = 25°C$.*

50

6.3 Tuning of the individual lasers

The previous sections showed that wider wavelength tuning is obtained when connecting the individual heaters in series, which will be the configuration used in the next sections. In addition, single-mode operation is maintained during the wavelength tuning, with a reproducibility of the order of the mode spacing. This makes these lasers (and heaters) interesting for many applications. This being said, a trade off must be made between the tuning range and the output power drop, as the wider the wavelength tuning the stronger the power drop.

6.3.1 DBR-RW reference laser tuning

In a similar order to the previous chapters, the first laser that will be considered is the DBR-RW reference laser. The emission wavelength as function of the heater current is shown in Fig. 6.7(a). Once again, a quadratic tuning behaviour is observed with single-mode operation maintained over a tuning range of about 4.4 nm.

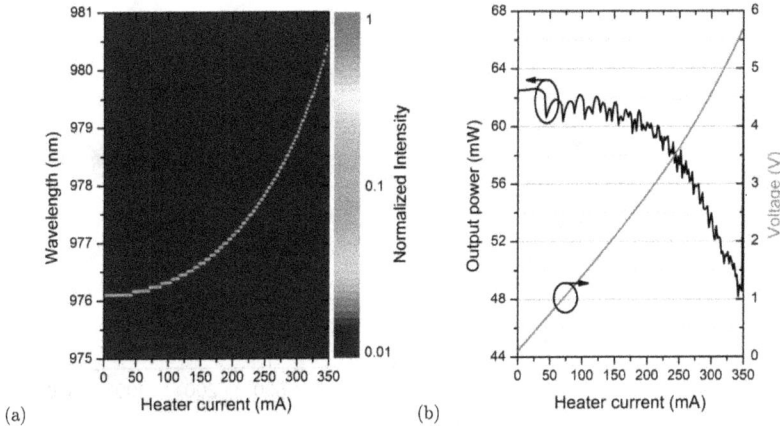

Figure 6.7: *Heater characteristics showing (a) the emission wavelength and (b) the emitted output power and heater voltage as function of the heater current at $T = 25°C$.*

Fig. 6.7(b) shows the corresponding output power and heater voltage drop during the wavelength tuning, indicating a power drop of about 7% (SDOM). From the power measurement, it can be seen that the voltage does not increase linearly with current, which indicates that the resistor is slightly temperature dependent; $R(T)$. Note that the reference laser has a 1000 μm long DBR grating, in comparison to the 500 μm long DBR gratings of the Y-branch lasers. In addition, the observed individual kinks and ticks in Fig. 6.7(b) correspond to mode hops as indicated in Fig. 6.7(a).

6.3.2 Y-branch DBR-RW laser tuning

The three Y-branch DBR-RW lasers will be considered next, namely the one based on a Sin, Cos and the SB structures. Fig. 6.8 shows the wavelength tuning as function of the heater current, where single-mode operation is observed from all three lasers with a $\Delta\lambda = 3.9$ nm tuning range. A total tuning coverage of 6.2 nm between 973 to 979 nm is obtained with a 2.3 nm grating spacing. With proper grating wavelength spacing, a total tuning of 2×3.9 nm = 7.8 nm can be achieved from these lasers.

The corresponding output power variation during the wavelength tuning is seen in Fig. 6.9. In this case, an output power variation (SDOM) less than 2% is observed.

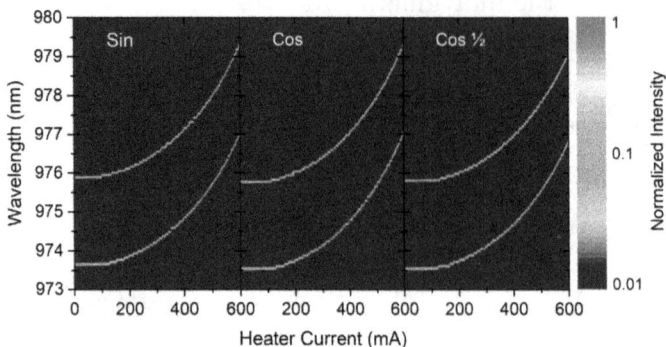

Figure 6.8: *False color contour plot of the spectral tuning of the emission wavelength of the Y-branch DBR-RW lasers (superimposed) as function of the heater current at $T = 25°C$.*

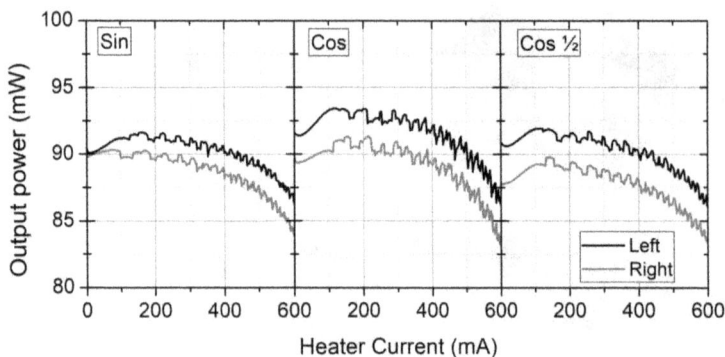

Figure 6.9: *Output power and heater voltage of the Y-branch DBR-RW lasers as function of the heater current at $T = 25°C$.*

6.3.3 Four-arm DBR-RW laser tuning

The tuning of the four-arm lasers is also investigated, where each laser has four micro-heaters embedded on top of each DBR grating, see Appendix E for a photograph of this device. These four heaters are connected in series as discussed in the previous section. Fig. 6.10 shows the emission wavelength as function of the heater current where 2.7 nm tuning is obtained from each arm. Note that this discrepancy with the Y-branch lasers ($\Delta\lambda = 3.9$ nm), is explained by the 850 μm long gratings, compared to the shorter 500 μm gratings of the Y-branch lasers. In addition, the larger heater dimensions as well as chip size corresponds to a larger heating surface.

In the case of Laser A (with a single intersection point), the inner arms shows some regions with multi-mode operation (< 80 pm at FWHM), which is still smaller than the targeted applications. The outer arms of Laser A mainly show single-mode operation with a spectral linewidth of 17 pm. In the case of laser B (with two intersection points), single-mode operation is observed from all four arms. This being said, B1 and B2 show a fall-back in the wavelength around a heater current of 475 mA, after which the wavelength increases once again. In both cases, the four-arm lasers covers a tuning range of about 9 nm from about 971 to 980 nm. With proper grating design, 4×2.7 nm $= 10.8$ nm of tuning can be obtained.

The corresponding power and heater voltage during the wavelength tuning is seen in Fig. 6.11. A small power drop of about 3% is observed for both lasers, with overall similar power and voltage characteristics between the individual arms.

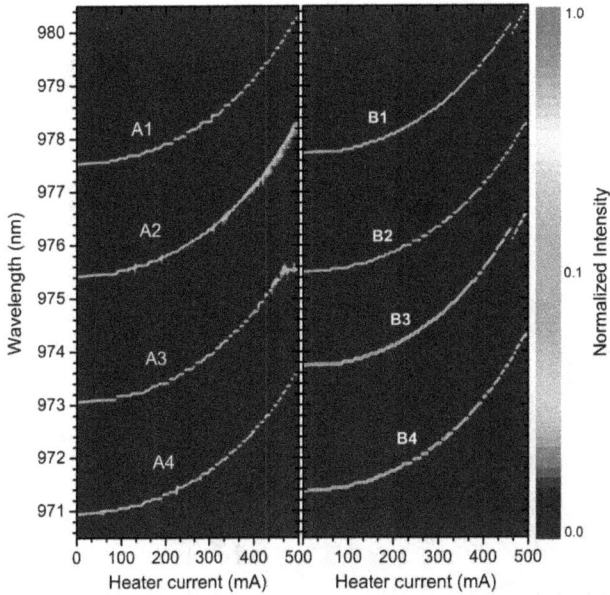

Figure 6.10: *Emission wavelengths of the individual arms of the two four-arm lasers, as function of the heater current at $T = 25°C$.*

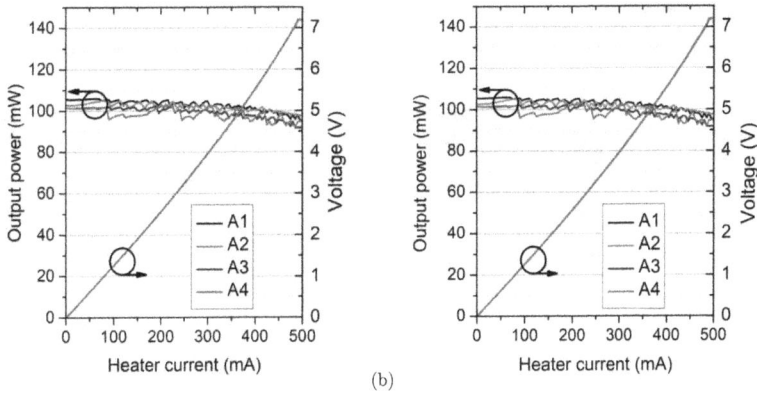

(a)   (b)

Figure 6.11: *Output power and heater voltage of laser A and B, as function of the heater current at $T = 25°C$.*

6.3.4 Six-arm DBR-RW laser tuning

Finally, the same investigation is made for the six-arm laser, which has six individual micro-heaters embedded on top of its DBR gratings. A photograph of this device can be found in Appendix E. Fig. 6.12(a) shows the emission wavelength of the individual arms as function of the heater current, where each arm can be tuned by 2.9 nm, providing in total more than 9 nm of combined tuning. As can be seen, each of the six arms has regions of multi-mode operation, in particular above a heater current of 400 mA. This multi-mode operation was also observed earlier in the electro-optical investigation, see Sec. 5.2, and is therefore not a feature of the micro-heaters.

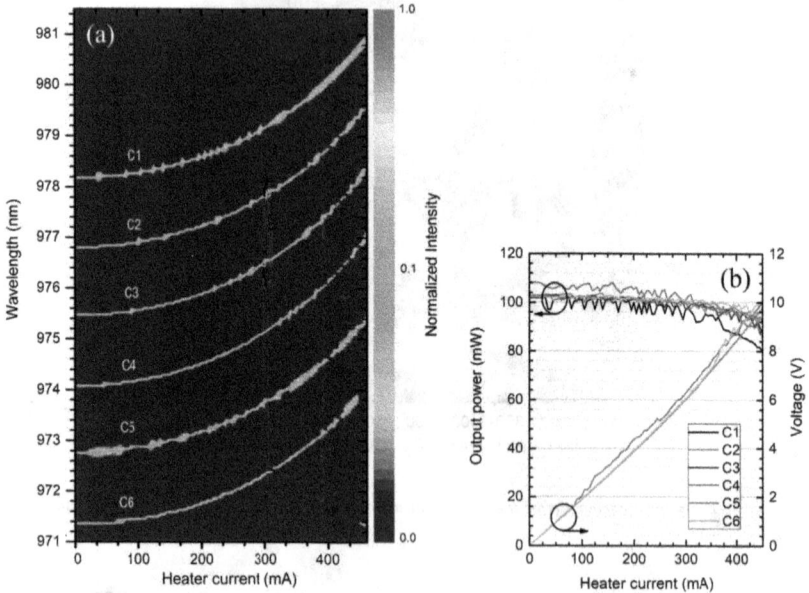

Figure 6.12: *Heater characteristics showing (a) the emission wavelength and (b) the emitted output power of the six-arm laser, as function of the heater current at $T = 25°C$.*

The corresponding output power and heater voltage characteristics are seen in Fig. 6.12(b), where an output power drop of up to 7% is seen. The combination of the six heaters provides a higher resistance, which explains the high voltage drop of about 10 V reached at a heater current of 450 mA. The tuning was stopped around a heater current of 450 mA due to a 10 V limit of the applied power supply.

6.4 Conclusion

In summary, the tuning capabilities of the micro-heaters have been presented in this chapter. One of the main findings is the extended tuning range obtained when connecting multiple micro-heaters in series. This configuration provides the highest induced electrical heater power, and subsequently the widest wavelength tuning. On the other hand, due to this induced large thermal gradient, this configuration also provides the strongest output power drop during the wavelength tuning.

The reproducibility of the micro-heaters was also investigated, which proved over multiple measurements a SDOM below of 0.5 mW in output power and below 30 pm in emission wavelength. Note that the latter value is below the mode spacing, estimated to be about $\delta\lambda = 52$ pm, and about two times the 17 pm resolution limit of the spectrometer.

The tuning characteristics of the individual lasers are summarized in Table 6.2. It was shown that single-mode operation is maintained during the wavelength for all lasers, beside the Six-arm DBR-RW laser which had regions of multi-mode emission. This is however due to the laser cavity itself, as was shown in Sec. 5.2, and not due to the tuning of the micro-heaters.

Table 6.2: *Tuning characteristics of a single arm of the different investigated lasers.*

Laser (single arm)	I_{heater} [mA]	U_{heater} [V]	P_{heater} [W]	$\Delta\lambda$ [nm]	ΔT [K]	R_{heater} [Ω]	ΔP [%]
DBR-RW	350	5.70	1.99	4.4	67.5	16.2	6.6
Y-branch	600	3.79	2.27	3.9	60.5	6.3	< 2
Four-arm	500	7.19	3.59	2.7	42.0	14.4	< 3
Six-arm	460	9.99	4.59	2.9	44.5	21.7	< 7

The findings of this investigation support the motivation behind the use of multi-wavelength lasers. By proper distancing the wavelength of each DBR grating, such that one begins where the tuning of the previous grating ends, an extended tuning of $\Delta\lambda = C \; N_{arms}$ can be realized. Here, C denotes the tuning range of the applied micro-heater for a laser of N_{arms} arms. E.g. for the six-arm laser, a total tuning range of $\Delta\lambda = 6 \times 2.7$ nm $= 17.4$ nm can be realized. The tuning range C can also be increased by optimizing the micro-heater design, by e.g. considering its' and the DBR dimensions. The larger the heater surface the higher induced heater power, and the smaller the DBR grating the higher induced temperature gradient and thus wider tuning. By using 500 μm long DBR gratings (as in the Y-branch lasers), the six-arm laser can potentially provide 6×3.9 nm $= 23.4$ nm of combined tuning.

Different parameters can be considered when optimizing such heater elements, such as the dimensions of the micro-heaters, DBR gratings as well as the chip dimensions. The induced heating should optimized to the grating structures, and the heating of the surroundings should be minimized as it is mainly lost heat. In addition, this extra heating which does not contribute to the wavelength tuning, heats up the entire chip resulting in a stronger power drop. The different heater, grating and chip dimensions explains the different tuning performance of the different lasers in Table 6.2. That is e.g. why the four- and six-arm lasers have narrower tuning $\Delta\lambda$ than the Y-branch and DBR-RW lasers, although larger heater powers P_{heater} are induced.

The upper tuning range limit would be the width of the gain profile, which depends on the QW structure of the active region. One way of extending the gain profile, e.g. by implementing MQWs with varying thickness and/or compositions in a single active region [85, 86, 87]. This approach is used in superlumi-nescent laser diodes [88]. The drawback however is the reduced output power, as the gain is "flattens out" over a wide spectral range.

The speed at which the tuning can take place was not presented in this study, although simple tests in the Lab indicates that 4.4 nm of tuning can be obtained within tens of milliseconds. This is ultimately limited by the material (GaAs) heat dissipation rate, i.e. how quick the heating/cooling of the material can take place.

Finally and as already mentioned, a trade-off must be made between the tuning range and drop in the output power. This is crucial in e.g. MOPA applications, where a change in the seed power can affects the output power of the PA. This will be investigated in more details in chapter 8.

Chapter 7

Sampled-grating DBR-RW laser

SG based lasers can provide wide wavelength tuning ($>$ 100 nm), and are well established in InP for optical telecommunication [6, 89]. In particular, these light sources are used within the field of dense wavelength division multiplexed (DWDM) fiber-optic networks [7, 8, 9].

While InP based SG lasers are well established, this work (to the authors best knowledge) is the first demonstration of GaAs based SG lasers. The challenges in realizing such light sources are explained by the GaAs technological and material-related difficulties, mainly in the realization and integration of the different elements. This includes the shorter wavelength of GaAs which requires smaller grating periods, the difficult regrowth of GaAs/AlGaAs which requires high temperature to avoid bulk oxygen and the risk of surface oxidation [90].

In addition, electrical tuning is challenging in GaAs due to low non-radiative carrier lifetime of $<$ 5 ns, and electrical tuning of only -0.3 nm has been demonstrated [64]. As a result, thermal tuning is used to control the wavelength of the presented GaAs lasers.

In this chapter, the concept and design of SG lasers is presented, including numerical simulations describing the performance of the designed SG lasers. This is followed by a description of the development process of GaAs based vertical structures. Experimental investigations are then carried out on different types of developed SG structures. Finally, SG-DBR lasers emitting around 976 nm are experimentally characterized and their results are compared with the numerical simulations.

7.1 Concept and design of SG lasers

A schematic of a typical SG-DBR laser is shown in Fig. 7.1. The cavity is defined by a gain section, a phase section and two SG sections serving as the front and back mirrors (FM and BM, respectively). A SG consist of a periodic sequence Λ_s of regions with and without a Bragg grating, of length L_g and $\Lambda_s - L_g$, respectively. Such structures provide comb-shaped reflection spectra, which are the basis for the tuning mechanism, see Fig. 7.2. Lasing occurs at the overlap between the two reflectivity spectra of the FM and BM i.e. at $\lambda_0 = 976$ nm.

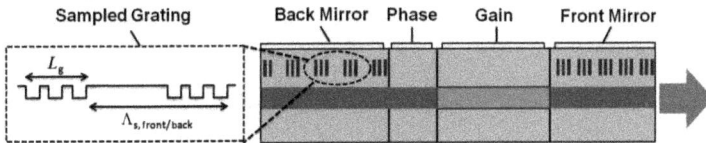

Figure 7.1: *Schematic of a SG-DBR laser consisting of a gain section, a phase section, a FM and a BM. The red sections represent the passive waveguide, whereas the blue represent the active section where the gain material is embedded. Zoom: A SG consists of a periodic sequence of regions with (L_g) and without a Bragg grating ($\Lambda_s - L_g$).*

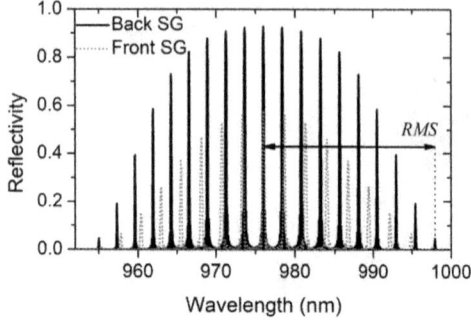

Figure 7.2: *Simulated reflectivity curves of the FM (black), the BM (red) and the corresponding RMS.*

Wavelength tuning is obtained when the reflectivity spectrum of one/both SGs are tuned, which can either be done electrically or thermally. In the latter case and as will be employed in this thesis, a change in temperature will induce a change in the effective index of refraction Δn, which provides a tuning of the corresponding reflectivity spectra. As described in chapter 6, the induced heat shifts the reflectivity spectrum towards longer wavelengths.

The active section of the laser lies within the cavity, typically consisting of embedded SQW or MQW. The structure also includes a section for phase control, which can be used to adjust the position of the longitudinal modes relative to the reflectivity spectra of the two mirrors, and hence improve the SMSR.

The period Λ_s defines the peak-peak distance between the reflectivity peaks according to the relation [91]

$$\Delta\lambda_s = \frac{\lambda_0^2}{2n_g\Lambda_s} \ . \tag{7.1}$$

As an example and as will be used for the BM; a period of $\Lambda_{s,back} = 50$ μm, $\lambda_0 = 976$ nm and a group index of $n_g = 4.0$ (experimentally estimated for GaAs [92]), provides a peak-peak distance of $\Delta\lambda_{s,back} = 2.38$ nm. This means that the temperature of the corresponding mirror section must at least be changed by 37 K, for a tuning coefficient $\Delta\lambda/\Delta T = 6.5$ pm/K (see Sec. 6.1), to realize the full tuning potential of the laser.

The reflectivity spectra of three examples of period lengths: $\Lambda_s = [50, 100, 150]$ μm are shown in Fig. 7.3, which provide peak-peak distances of $\Delta\lambda_s = [2.4, 1.2, 0.8]$ nm, respectively. These peak-peak distances can easily be obtained using micro-heaters as demonstrated in chapter 6.

Similar to a Vernier tool, the tuning mechanism of a SG-DBR laser is based on the period difference between the FM and the BM; $\Delta\Lambda_s = \Lambda_{s,back} - \Lambda_{s,front}$ [41]. By having different periods, different peak-peak distances are obtained, however with both being centred around $\lambda_0 = 976$ nm. The two reflectivity curves will then coincide once again at a distance called the Repeat-Mode-Spacing (*RMS*), which describes the total possible tuning, see Fig. 7.2. The RMS can be approximated by [91]

$$RMS \approx \frac{\lambda_0^2}{2n_g\Delta\Lambda_s} = \frac{\lambda_0^2}{2n_g\left(\Lambda_{s,back} - \Lambda_{s,front}\right)} \ . \tag{7.2}$$

By combining a $\Lambda_{s,back} = 50$ μm with a $\Lambda_{s,back} = 45$ μm so that $\Delta\Lambda_s = 5$ μm, a $RMS \approx 23.8$ nm can be obtained, see Fig. 7.2. Eq. (7.2) also indicates that an increased RMS and hence wider tuning can be achieved with a smaller period difference $\Delta\Lambda_s$.

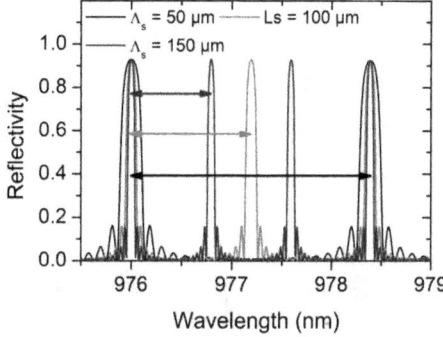

Figure 7.3: *Reflectivity spectra around $\lambda_0 = 976$ nm for periods of length $\Lambda_s = [50, 100, 150]$* μm.

7.1.1 Passive cavity simulations

The shown reflectivity spectra in Fig. 7.2 and 7.3 were obtained using the Transfer matrix model defined as

$$\begin{bmatrix} a(z) \\ b(z) \end{bmatrix} = \mathbf{T}(z) \begin{bmatrix} a(0) \\ b(0) \end{bmatrix} . \tag{7.3}$$

This model describes the transfer of a forward and a backward propagating field amplitudes a and b, respectively, from a position $z = 0$ to an arbitrary position z after passing a medium described by $\mathbf{T}(z)$. The Transfer matrix of a dielectric section containing a Bragg grating is given by [93]

$$\mathbf{T}(z) = \begin{bmatrix} \cos(\gamma z) - i\Delta\beta_p \frac{\sin(\gamma z)}{\gamma} & -i\kappa^+ \frac{\sin(\gamma z)}{\gamma} \\ i\kappa^- \frac{\sin(\gamma z)}{\gamma} & \cos(\gamma z) + i\Delta\beta_p \frac{\sin(\gamma z)}{\gamma} \end{bmatrix} . \tag{7.4}$$

The parameter $\gamma = \sqrt{(\Delta\beta_p)^2 - \kappa^+\kappa^-}$ includes the passive propagation factor $\Delta\beta_p$ and the coupling coefficients κ^+ and κ^- for the forward and backward propagating waves, respectively. The passive propagation factor is defined as

$$\Delta\beta_p(z,\lambda) = \beta(z,\lambda) - \beta_0 = \frac{2\pi}{\lambda}n(\lambda) - \frac{i}{2}\alpha_0 - \frac{2\pi}{\lambda_0}n_0 , \tag{7.5}$$

where $n(\lambda)$ is the modal index of refraction, α_0 the modal absorption coefficient, $\lambda_0 = 976$ nm the center wavelength and n_0 is the reference effective index. The coupling coefficients of the grating defined as $\kappa^\pm = \kappa e^{\mp 2i\pi\phi}$ with ϕ being the grating phase, depends on the etching depth, width and the order of the grating. In sections without a Bragg grating, these coefficients are simply set $\kappa^\pm = 0$.

The Transfer matrix $\mathbf{T}(z)$ in eq. (7.4) describes a single dielectric section, and the corresponding matrix for a stack of sections of lengths $L_1, L_2, ..., L_n$ is simply obtained by the multiplication

$$\mathbf{T}(L) = \mathbf{T}_n(L_n)\mathbf{T}_{n-1}(L_{n-1})\cdots\mathbf{T}_2(L_2)\mathbf{T}_1(L_1) \quad \text{for} \quad L = \sum_{i=1}^{n} L_i . \tag{7.6}$$

In the particular case of a SG, the corresponding $\mathbf{T}(z)$ matrix can be simplified as

$$\mathbf{T}_{\mathrm{SG}}(L) = (\mathbf{T}(L_g, \kappa)\mathbf{T}(\Lambda_s - L_g, \kappa = 0))^m , \tag{7.7}$$

which contains a grating section and a sampling section (without a grating) repeated m number of times, i.e. number of periods. Finally, the overall reflectivity coefficient of the stack is obtained by the elements

59

of the Transfer matrix as $r_L = T_{12}(L)/T_{22}(L)$. This model was implemented into a MATLAB script which can be found in Appendix F.

In order to achieve high reflectivity from a SG, a structure with a high coupling coefficient is needed, as the grating only fills a part of the total mirror section.

This effect is simulated for coupling coefficients of; $\kappa = 200$ cm^{-1} and $\kappa = 400$ cm^{-1}, and the corresponding maximum reflectivities (at λ_0) are shown in Fig. 7.4. As can be seen, the reflectivity increases with increasing coupling coefficient κ, grating length L_g and with larger number of sampling periods m, and vice versa. The first step of designing a SG-DBR laser is choosing the appropriate SG structures for the FM and BM.

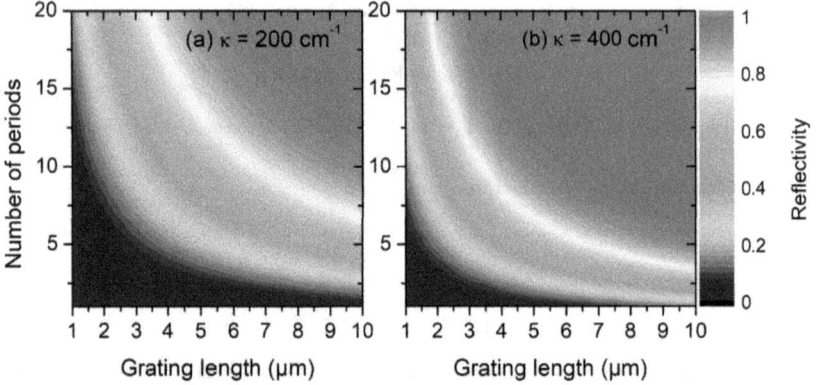

Figure 7.4: *Simulation of the maximum reflectivity of a SG having a coupling coefficient of (a)* $\kappa = 200$ cm^{-1} *and (b)* $\kappa = 400$ cm^{-1}, *as function of the grating length* L_g *and number of periods* m.

In the shown example of Fig. 7.2, the BM consist of $m = 20$ periods of $\Lambda_{s,back} = 50$ μm with a grating length $L_g = 5$ μm providing a maximum BM reflectivity $R_{back,max} \approx 90\%$ for $\kappa = 200$ cm^{-1}. In most lasers, the FM should have lower reflectivity and in the shown example in Fig. 7.2, the FM consists of $m = 10$ periods of $\Lambda_{s,front} = 45$ μm with the same grating length of $L_g = 5$ μm, providing $R_{front,max} \approx 58\%$.

Alternative models to the Transfer matrix model exist, which describe SG lasers with analytical expressions. One of these models by Jayaraman et al. [91] is considered in Appendix G and compared to the Transfer matrix model. Based on the limitations of the analytical model, the Transfer matrix model was used in this work.

7.1.2 Longitudinal mode spacing

In this section the active cavity is considered, in particular the length of the gain section. The mode spacing inside a SG cavity can be described as

$$\delta\lambda = \frac{\lambda_0^2}{2n_g L_{Cavity}} = \frac{\lambda_0^2}{2n_g \left(L_{front,eff} + L_{gain} + L_{phase} + L_{back,eff}\right)} . \tag{7.8}$$

In the case of a SG, the cavity consists of the gain and phase sections, plus an effective penetration depth into the FM and BM. The effective length of a grating can be obtained from the Transfer matrix model, and is further described in Appendix C. From eq. (7.8), it is clear that a larger mode spacing is obtained for a shorter cavity.

The impact of the mode spacing and the cavity length is illustrated in Fig. 7.5 which shows the longitudinal mode spacings of (a) $L_{gain} = 900$ μm and (b) $L_{gain} = 1900$ μm, both having a $L_{phase} = 100$ μm and the before mentioned FM and BM specifications. In the case of $L_{\text{gain}} = 900$ μm, a mode spacing of $\delta\lambda = 90$ pm is obtained, while the longer gain section of $L_{\text{gain}} = 1900$ μm gives a mode spacing of $\delta\lambda = 50$ pm. Note that the reflectivity product FM × BM is also shown in Fig. 7.5 as it represent the reflectivity seen inside the cavity.

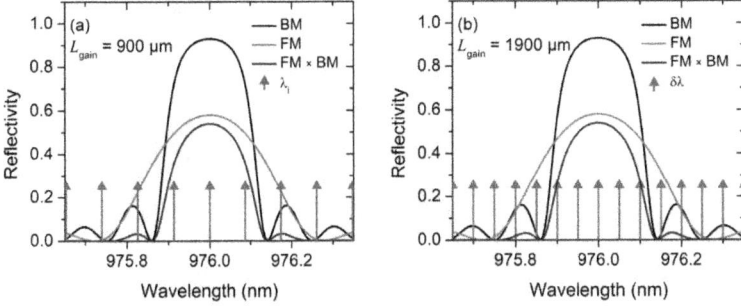

Figure 7.5: *Longitudinal modes λ_i for a cavity of gain length (a) $L_{\text{gain}} = 900$ μm and (b) $L_{\text{gain}} = 1900$ μm, next to the reflectivity curves of the FM, BM and the product FM × BM.*

In the case of $L_{\text{gain}} = 900$ μm, few (three) longitudinal modes fits within the reflectivity product of the FM/BM, and thus single mode operation with a relatively good SMSR can be expected. In the case of $L_{\text{gain}} = 1900$ μm gain, multiple longitudinal modes fits within the reflectivity curves, and thus worse SMSR can be expected from such a cavity.

Reducing the gain length even further, say $L_{\text{gain}} = 400$ μm, will provide a longitudinal mode spacing of about $\delta\lambda = 136$ pm. In this case, only a single longitudinal mode will fit within the reflectivity curves, and an even better SMSR can be expected. This being said, shorter gain sections provide less output powers and thus a trade-off must be made.

7.1.3 Active cavity simulations

In this section, active cavity simulations are carried out to investigate the dynamics of the studied SG-DBR laser during wavelength tuning. This study is based on the coupled wave equations [94]:

$$\begin{bmatrix} \frac{\partial}{\partial z} + i\Delta\beta_a(z) & i\kappa^+(z) \\ i\kappa^-(z) & -\frac{\partial}{\partial z} + i\Delta\beta_a(z) \end{bmatrix} \begin{bmatrix} a(z) \\ b(z) \end{bmatrix} = 0 , \tag{7.9}$$

with the active propagation factor

$$\Delta\beta_a = \frac{2\pi}{\lambda_0}\Delta n + \frac{n_g}{c}\Omega + \frac{i}{2}[g - \alpha_0] . \tag{7.10}$$

In the above expression, c is the speed of light in vacuum, Δn is the difference between the sectional effective index of refraction and the reference index (set to 0 here), g is the modal gain of the active material and $\alpha_0 = 3$ cm^{-1} is the modal absorption coefficient [92]. The complex frequency (eigenvalue) Ω is a solution of $b(L) - r_L a(L) = 0$ for the initial condition $a(0) = r_0 b(0)$ as described in [95].

From this frequency, the threshold gain g_{th} can be obtained as a solution of $\text{Im}\,(\Omega(g_{\text{th}})) = 0$. The gain margin Δg of the cavity modes needed to reach threshold is obtained as

$$\Delta g \approx \frac{2n_g}{c}\,\text{Im}(\Omega)\,\frac{L_{\text{total}}}{L_{\text{gain}}} , \tag{7.11}$$

and the corresponding wavelengths of the cavity modes are given by

$$\Delta\lambda = \frac{d\lambda}{d\omega}\,\text{Re}(\Omega) . \tag{7.12}$$

The model above is implemented in the *cme* software developed at the Ferdinand-Braun-Institut, which can simulate the active cavity of SG-DBR lasers near threshold [96].

61

7.1.4 Tuning of a single SG

The simplest wavelength tuning scheme of a SG laser would be the tuning of one of the two mirrors, resulting in a step-like tuning behaviour. This is seen in Fig. 7.6 for the case of $L_{gain} = 900$ μm, where the black curve shows the wavelength change $\Delta\lambda$ during the index changing (tuning) Δn of the FM. This results in a discrete wavelength tuning from -9.5 to $+9.6$ nm, hence providing a total tuning of about 19 nm. The discrete tuning "steps" are spaced by 2.3 nm, in accordance with the FM peak-peak design ($\Lambda_{s,front} = 45$ μm $\Rightarrow \Delta\lambda_{s,front} = 2.3$ nm).

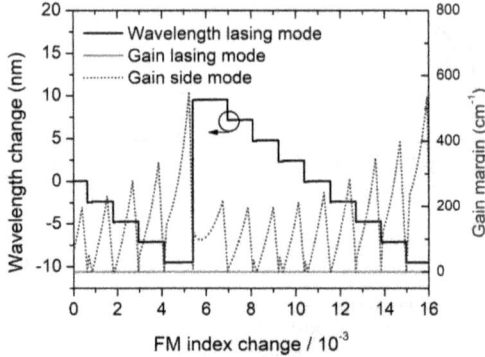

Figure 7.6: *Wavelength change of the lasing mode (left axis) and the gain margin of the main and side-modes (right axis), as function of the index change of the FM.*

Fig. 7.6 also shows the gain margin Δg of the lasing and of the side-mode. The gain margin for the lasing mode is per definition 0 cm^{-1}, while it varies for the side-mode during the wavelength tuning. It is seen that the gain margin of the side-mode has a maximum at ~ 550 cm^{-1}, and a minimum (goes to 0 cm^{-1}) at each of the edges of the discrete wavelength steps. The latter corresponds to the case where there will be a competition between the main and the side-mode. In the presentation of Fig. 7.5, this corresponds to having two longitudinal modes symmetrically positioned around the reflectivity peaks.

In addition, local maxima (and few minimas) are seen in Fig. 7.6, suggesting an optimum (and pessimum) operation setting at each discrete wavelength step. This implies that single mode operation with high SMSR can be found at these maxima, as the "main mode" would be centered around R_{peak}, and vice versa for the local minima [97].

In the case of BM index change, the wavelength tuning also follows a stair-like behaviour, however with opposite sign (increasing with the index change) to the one shown in Fig. 7.6. The wavelength is then tuned between $+10.7$ to -8.0 nm, covering an ~ 19 nm tuning range. The gain margin of the side-mode during the BM index change ranges between 0 and 580 cm^{-1}, in a similar fashion to the case of FM index change.

Finally, the effect of the mode spacing is once again shown, but in terms of the gain margin. Fig. 7.7 shows the gain margin for the case of $L_{gain} = 900$ μm (same as in Fig. 7.6) next to the case of $L_{gain} = 1900$ μm, which has a maximum value of about 238 cm^{-1}. The larger gain margin is obtained for the shorter cavity, which corresponds to less competition between the longitudinal modes compared to the longer cavity, and suggests improved SMSR. This is in agreement with the description in Sec. 7.1.2.

7.1.5 Tuning of both SGs

For a continuous wavelength tuning, both mirrors must be tuned simultaneously. Fig. 7.8 shows a tuning map of the studied laser ($L_{gain} = 900$ μm) under the index change Δn of both the FM and BM. When both mirrors are tuned simultaneously and equally, a diagonal "tuning line" is obtained, where the wavelength will increase linearly from λ_0 (1) to $\lambda_0 + 2.4$ nm (2), see Fig. 7.8. Such a tuning is similar

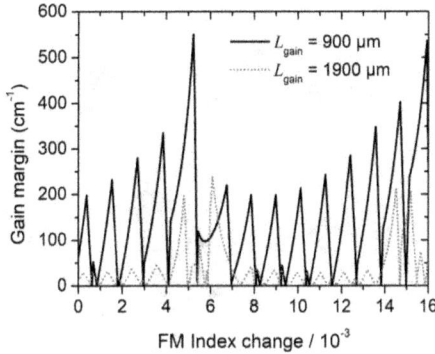

Figure 7.7: *Gain margin of the side-mode as function of the index change of the FM for* $L_{\text{gain}} = 900$ *μm (black curve) and for* $L_{\text{gain}} = 1900$ *μm (red curve).*

to the thermal tuning of DBR (see Fig. 6.1 in chapter 6) and of DFB lasers [40]. If a further continuous tuning is desired, one would have to go to the next diagonal line; +2.4 nm (3) to +4.8 nm (4), (5), and the next line, until passing every diagonal tuning-line and hence cover the whole tuning range. Fig. 7.9 shows the tuning map in a 3D presentation, indicating the slope of the individual regions and how each region ends at a wavelength where the next begins. The wavelength can be tuned between −10.9 nm up to +11.3 nm, thus covering a 22.2 nm tuning range, which is close to the estimated RMS value of 23.8 nm obtained from eq. (7.2).

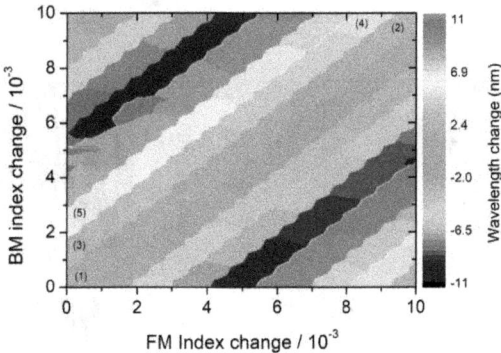

Figure 7.8: *Wavelength tuning-map as function of the index change* Δn *of the FM and BM, respectively.*

7.1.6   Tuning of the phase section

The effect of the phase section is considered next in Fig. 7.6, which shows (a) the wavelength change and (b) the gain margin, as function of the index change Δn of the phase and of the FM.

The results indicate that the phase section does not affect the lasing wavelength, however it influences the gain margin. The gain margin indicates that there would be an optimum (pessimum) operation setting, which provides the highest (lowest) SMSR. This optimum position corresponds to the situation where a

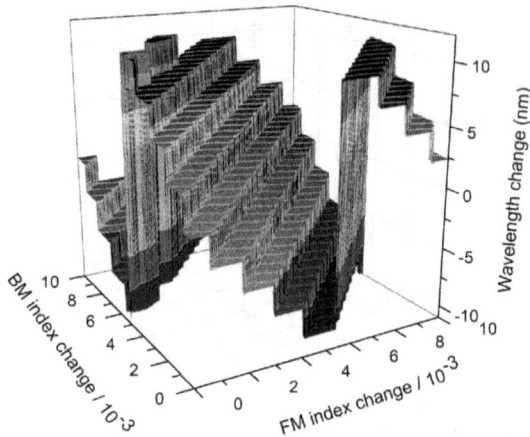

Figure 7.9: *3D representation of the wavelength change as function of the index change Δn of the FM and BM, respectively.*

single longitudinal mode is centred at the reflectivity curve maxima, see Fig. 7.5(a). The pessimum case corresponds to having two longitudinal modes centred within the reflectivity curves.

An optimum operation setting can be found around $\Delta n \sim 35 \times 10^{-3}$ (and at $\Delta n \approx 83 \times 10^{-3}$), offering the largest gain margin throughout the wavelength tuning. This being said, the gain margin still goes to zero in between the discrete wavelength tuning steps (with increasing FM index) as seen in Fig. 7.6.

The FM index change in Fig. 7.10(a) causes a discrete wavelength tuning, as was already shown in Fig. 7.6

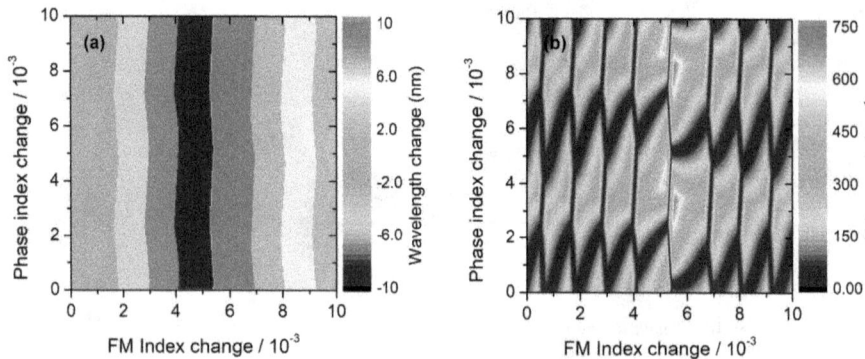

Figure 7.10: *Emission wavelength (a) and gain margin (b) as function of the index change Δn of the FM and of the phase section.*

64

7.2 Vertical structure

After describing the tools and design approach of SG-lasers, the vertical structure is now considered. A SG can be realized either as a surface or as a buried grating structure, however it is only in the latter case that high enough coupling coefficients can be realized. The developed vertical structure of the SG lasers in this work was grown in a two-step process by MOVPE on 3 inch GaAs wafers. The different steps of this process are illustrated in Fig. 7.11, including:

1. AlGaAs cladding layer grown on top of the GaAs substrate

2. Growth of the GaAs waveguide

3. Growth of the grating layer and an stop-etch layer

4. Growth of the InGaAs QW

5. Growth of the GaAs cap

6. Lithography step and selective etching (GaAs and InGaP) of the active area

7. E-beam lithography and selective wet etching to define the grating and waveguide areas

8. CBr_4-based shallow in-situ etching to remove the oxygen contamination

9. Second MOVPE growth of the p-side; GaAs waveguide, AlGaAs cladding and GaAs contact layer

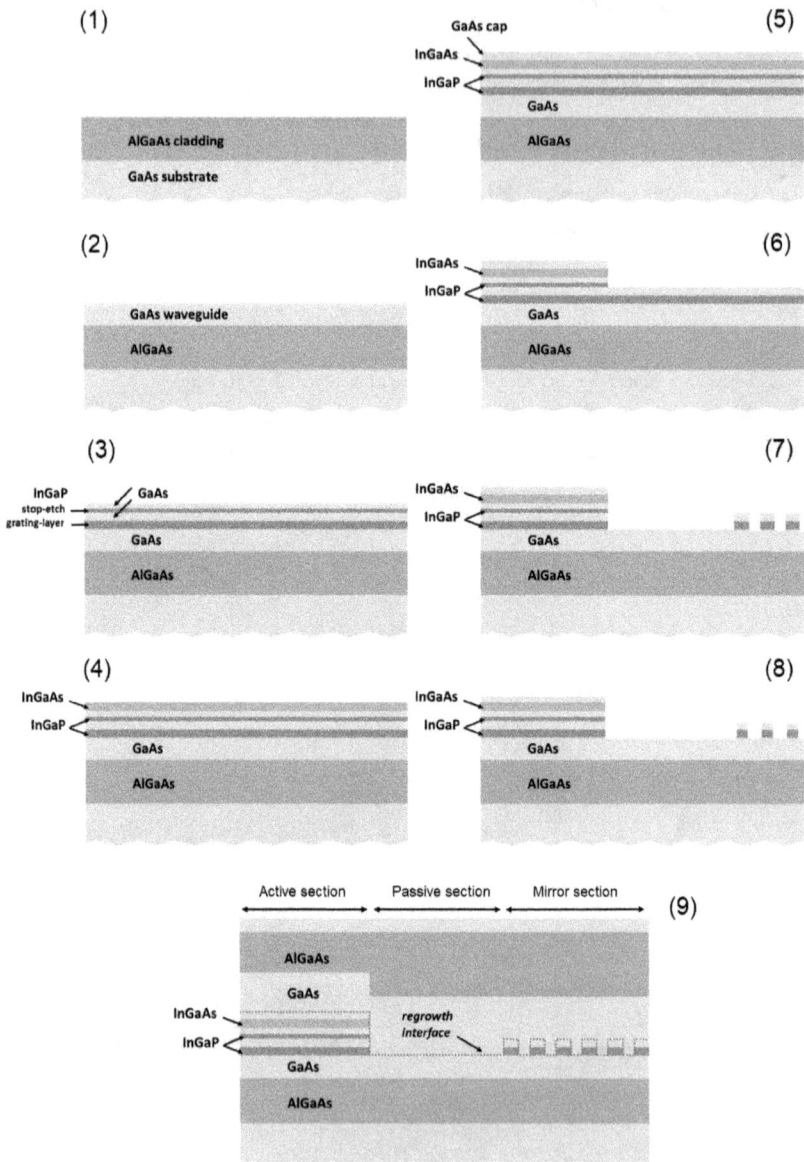

Figure 7.11: *Growth steps of the vertical structure used for the SG-DBR lasers.*

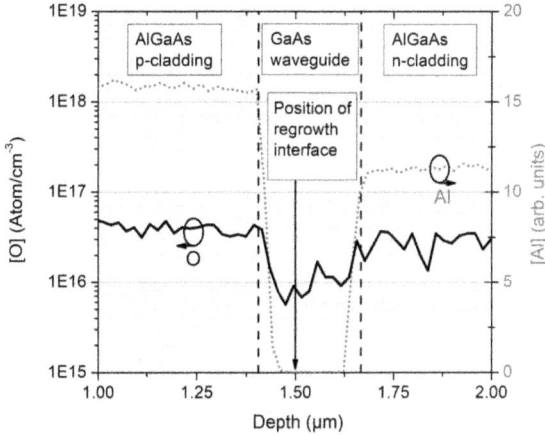

Figure 7.12: *Measured oxygen and aluminium concentration around the regrow interface.*

Between step (5) and (6) in Fig. 7.11, the GaAs cap layer was etched away until the InGaP stop-etch layer was reached. The InGaP was then etched away while using the GaAs below as a stop-etch layer.

After the wet etch in step (8) for the definition of the grating, a shallow etch of (nominally) 5 nm was done in-situ before the regrowth. This shallow etch reduces the thickness of the cap, nominally from 30 to 25 nm, and attacks the grating section both vertically and horizontally, with the result of shrinking a bit the protruding parts.

In Fig. 7.12 one can see the secondary ion mass spectrometry (SIMS) profiles of aluminium and oxygen. The aluminium profile has a through or depression corresponding to the waveguide. The regrowth interface position within this depression is indicated. This was done to remove contamination, in order to minimize losses and carrier recombination effects.

The transition between the gain and the grating sections is shown in Fig. 7.13, indicating a smooth transit between the layers, with some compositional perturbation in the AlGaAs layer and with a small interface reflectivity.

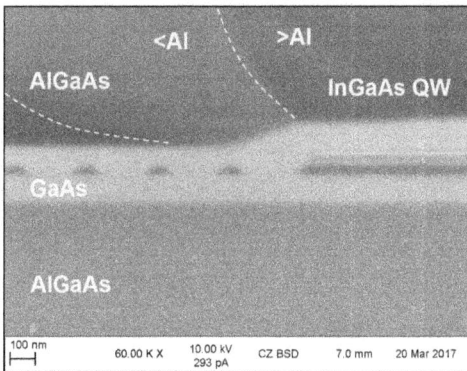

Figure 7.13: *SEM images of the transition between the gain and grating sections, obtained with backscattered electrons.*

The vertical structure, including its dimensions and material composition is illustrated in Fig. 7.14. As can be seen, two different structures were developed, one with a SQW and another with a double quantum well (DQW). Both structures utilize InGaAs QWs symmetrically sandwiched between GaAsP spacer layers, that were asymmetrically embedded in GaAs confinement and AlGaAs cladding layers. The n-confinement layer beneath the active region contains two InGaP layers acting as etch stop and grating layers, see Fig. 7.11. The placement of the grating layer near the active region ensures a high coupling coefficient κ.

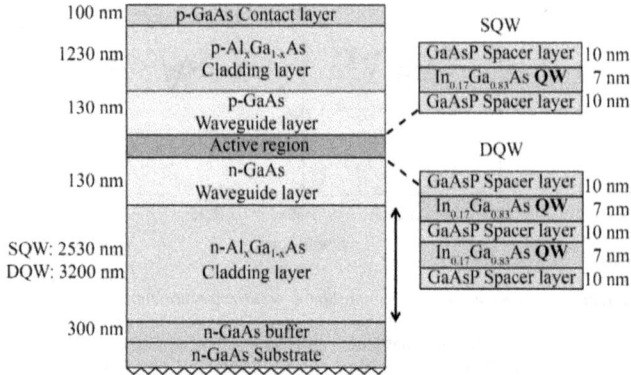

Figure 7.14: *Vertical structures of the SG lasers with a SQW or a DQW.*

Fig. 7.14 also shows two different n-cladding layer thickness's, which were used for the SQW and the DQW structures. The difference in composition and the data parameters are summarized in Table 7.1. In the case of the DQW, the larger n-cladding layer was grown to obtain a reduced vertical far field angle. This will be described in the next section, where the two vertical structures are experimentally characterized and compared.

Table 7.1: *Material data parameters of the developed vertical structures.*

Vertical structure	SQW	DQW
Thickness of n-$Al_xGa_{1-x}As$ (d_{cladding}) [μm]	2.53	3.20
Composition of n-$Al_xGa_{1-x}As$ (x)	0.25	0.17
Theoretical confinement factor Γ_{QW} [92]	0.015	0.022
Theoretical far field angle (FWHM) [92]	40°	28°

7.2.1 SG-design

The implemented SGs were based on the simulated design of Sec. 7.1. This includes grating bursts with grating lengths of $L_g = 5$ μm defined with Ebeam (Vistec SB251) in ma-N2401 resist. The bursts were repeated in periods of $\Lambda_{s,\text{front}} = 45$ μm and $\Lambda_{s,\text{back}} = 50$ μm for the front and back gratings, respectively. Within a burst, the gratings have a period of 143 nm (1st order) or 286 nm (2nd order), both with a duty cycle of 50%. The difference between the two order of gratings will also be considered in the next section. Fig. 7.15 shows the surface of the device over a SG section, where the position of the buried grating bursts can still be seen from the interference contrast, since the surface there is not perfectly planarized.

68

Figure 7.15: *Photograph of a SG, showing the individual grating burst and the implemented surface micro-heater with its bonding wires for electrical control.*

7.3 Grating characterization

Within the developed vertical structures of last section, grating structures were implemented which will be characterized next, with the focus on the difference in performance. In this section, these vertical structures are characterized, with focus on the difference in performance. This is done for the SQW / DQW based vertical structures having 1st or 2nd order gratings. This investigation was carried on devices consisting of an active section with a single grating section (either DBR or SG), see Fig. 7.16. These devices were in total 3000 μm long with 2.2 μm wide RW. At the back of these devices, the waveguide was tilted by an angle of $3°$ with respect to the normal, which suppresses reflections from the back facet to a power reflectivity $R_{back} < 10^{-3}$ [98]. Since no coating was applied to the front facets of the analyzed DBR/SG lasers, the power reflectivity is of the order of $R_{front} \approx 0.3$.

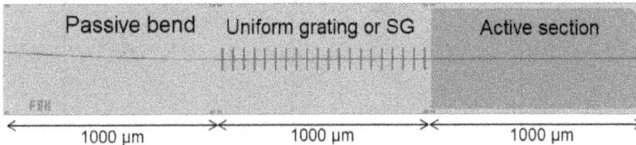

Figure 7.16: *Schematic of the analyzed devices.*

As described in the previous section, the vertical structures were grown with variations in the thickness and composition of the n-AlGaAs cladding layer, see Table 7.1.

The cross-section of the fabricated DBR lasers is shown in Fig. 7.17, obtained using an SEM. Image (a) shows a 1st order grating with a grating period of 143 nm while (b) shows the 2nd order grating (next to the active section) with a period of 286 nm, both having a duty cycle of 0.5. A smooth transition between the sections can be seen in the SEM images.

7.3.1 Electro-optical characteristics

The investigated devices were mounted p-side up on C-mounts and characterized at $T = 25°C$. The power characteristics are shown in Fig. 7.18 for the case of devices made on SQW / DQW based vertical structures with a 1st / 2nd order uniform Bragg gratings. Here, uniform grating refer to a DBR grating section, without interrupt or periods (as in a SG). The two SQW lasers have a threshold current of about $I_{SQW,th} = 23$ mA and slope efficiencies of $S_{SQW,1st} = 0.31$ W/A and $S_{SQW,2nd} = 0.29$ W/A for the 1st and 2nd order grating, respectively.

69

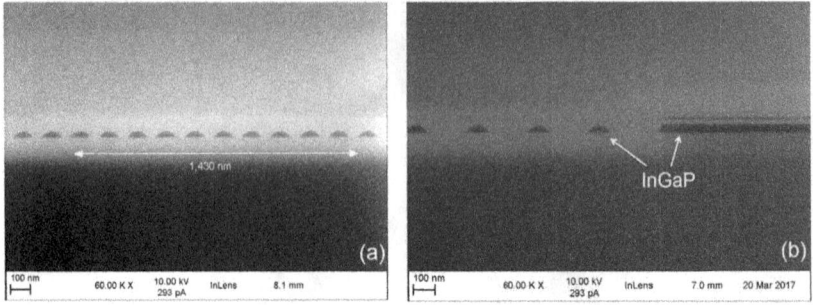

Figure 7.17: *SEM images of (a) 1st order and (b) 2nd order grating next to the active section.*

The DQW based lasers have a threshold current of $I_{DQW,th} = 30$ mA and slope efficiencies of $S_{DQW,1st} = 0.43$ W/A and $S_{DQW,2nd} = 0.40$ W/A, respectively. The SQW DBR lasers have lower threshold current and slope efficiency in comparison to the DQW lasers. The optical losses due to absorption are governed by free carriers originating from the ionized dopants ($\alpha_i \approx 0.5$ cm^{-1} [92]), which dominates over the absorption on carriers injected into the QWs. Therefore, the lower slope efficiency of the SQW structure is probably caused by an enhanced number of recombination centers in the confinement layers, resulting in an increased current leakage out of the QW above threshold. Since the slope efficiencies of the DBR with a 2nd order grating were reduced by only 7%, one can assume that the reflectivity of the Bragg reflector was only slightly decreased.

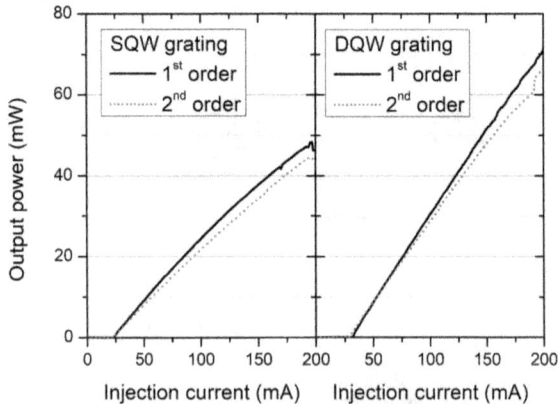

Figure 7.18: *Output power of DBR lasers with 1st or 2nd order uniform gratings developed on (left) SQW- and (right) DQW structure at $T = 25°C$.*

7.3.2 Far field characteristics

The measured far field profiles of the SQW and DQW structures are shown in Fig. 7.19 at $I_{\text{inj}} = 100$ mA. The DBR laser with a SQW structure has a vertical far field angle of 41°, while the DQW structure has a far field angle of 24°, both determined at FWHM. The slight asymmetry in the far fields is likely due to radiation effects into the substrate, indicating a too thin n-cladding layer.

Finally, the large far field angle of the SQW structure makes light collection a challenge, especially in miniatured optical systems (e.g. MOPA), where space and the number of optical components are limited.

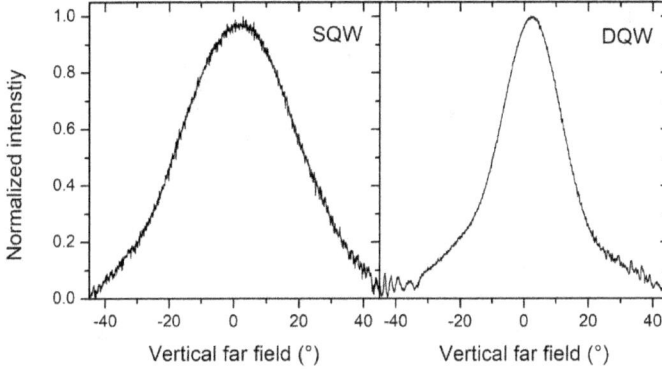

Figure 7.19: *Measured vertical far fields of the developed SQW and DQW structures at $I_{\text{inj}} = 100$ mA and $T = 25°C$.*

7.3.3 Optical characterization of the uniform DBR gratings

The optical properties of the developed vertical structures are studied next. Once again, this include devices processed on a SQW / DQW vertical structures having a 1st / 2nd order gratings.

In this section, the emission spectra of the devices with uniform DBR gratings were characterized by directly coupling the emitted light into a lensed optical fiber, connected to an optical spectrum analyzer (*Yokogawa AQ6370D*; resolution limit 20 pm).

The emission spectra of the SQW (left) and DQW (right) lasers with 1st (upper) and 2nd (lower) order DBR gratings for a sub-threshold current of $I_{\text{inj}} = 20$ mA are shown in Fig. 7.20. Longitudinal modes spectrally positioned within the Bragg reflection band show > 3 dB increased optical power, with respect to the background level of the amplified spontaneous emission (ASE).

The higher the coupling coefficient κ the broader the reflection band and the more longitudinal modes arise. The κ values were determined with a curve fitting based parameterized model as described in [95]. These values as well as the 3 dB widths of the Bragg reflection bands are summarized in Table 7.2.

Table 7.2: *Material data parameters of the developed vertical structure.*

Vertical structure	SQW		DQW	
Grating order N	1st	2nd	1st	2nd
Grating period Λ_r [nm]	143	286	143	286
Duty cycle D	0.5	0.25	0.5	0.25
Theoretical effective index n_{eff} [92]	3.408		3.424	
κ fitted from measured spectra [cm^{-1}]	497	292	251	152
Measured width of reflection band [nm]	4.0	2.4	1.8	1.4

71

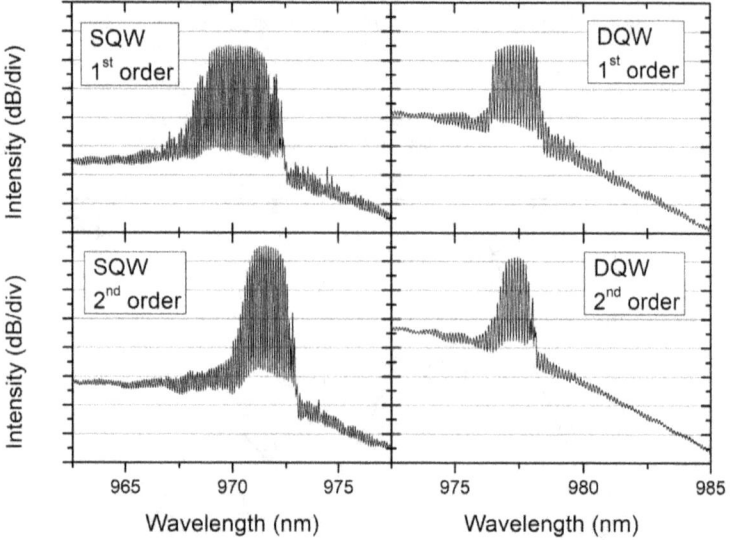

Figure 7.20: *Emission spectra below threshold* $(I_{\text{inj}} = 20 \text{ mA})$ *for DBR laser with uniform gratings (left) SQW- (right) DQW structure, (top) 1st and (bottom) 2nd order gratings at $T = 25°C$.*

In the case of the 1st order grating in the SQW structure, a coupling coefficient $\kappa > 497 \text{ cm}^{-1}$ was estimated. The 1st order grating in a DQW structure provides a coupling coefficient of $\kappa = 250 \text{ cm}^{-1}$. The center wavelengths of the reflection bands are around 970 nm and 977 nm for the SQW / DQW structures, respectively. These values correlates to the theoretically determined effective indices n_{eff} summarized in Table 7.2. These values were obtained according to the relationship

$$N\lambda = 2n_{\text{eff}}\Lambda_r , \qquad (7.13)$$

where N is the grating order and Λ_r is the grating period [99].

Widely tunable SG-DBR lasers require a broad gain spectrum to allow lasing of a sufficiently large number of SG modes. As can be seen in Fig. 7.20, the spectral width of the ASE (in good agreement with the gain dispersion) is only slightly affected by the number of QWs.

7.3.4 Optical characterization of the SGs

Similar optical characterizations were performed on devices with a SG instead of the uniform DBR grating. The investigated SGs consisted of $m = 10$ periods, with grating lengths of $L_g = 10 \text{ } \mu m$ and $\Lambda_s = 100 \text{ } \mu m$.

The optical spectra below threshold $(I_{\text{inj}} = 20 \text{ mA})$ are shown in Fig. 7.21, again for the case of SQW (left) and DQW (right) lasers with 1st (upper) and 2nd (lower) order gratings, respectively. All four spectra show 17 pronounced modes within a spectral range of about 20 nm. The distance between two SG modes is 1.2 nm, which is in a good agreement with eq. (7.1) $(\lambda_0 = 976 \text{ nm}, \Lambda = 100 \text{ } \mu m, n_g = 4.0 \Rightarrow \Delta\Lambda_s = 1.19 \text{ nm})$.

The sub-threshold spectra for the 1st and 2nd order gratings are comparable and the 2nd order gratings provide sufficient reflectivity to form SG modes. Pronounced modes can also be seen for the SG laser

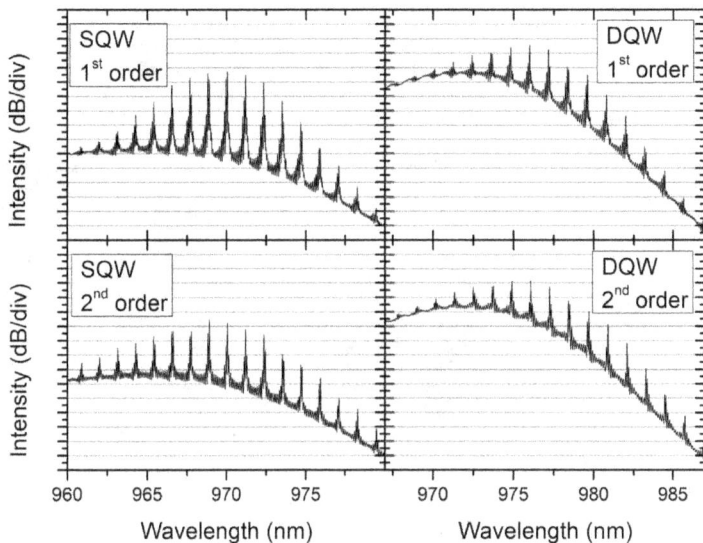

Figure 7.21: *Emission spectra below threshold* ($I_{\mathrm{inj}} = 20$ mA) *for SG lasers. (left) SQW- (right) DQW structure, (top) 1st and (bottom) 2nd order gratings at $T = 25°C$.*

with the DQW vertical structure despite the lower κ. The effect of the lower κ is seen in the lower peaks of the individual modes in Fig. 7.21 for the DQW structure in comparison to the SQW.

7.3.5 Summary

The results of this section show, that although the coupling coefficient κ of the DQW vertical structure is lower than the case of the SQW, a DQW structure can still be used in SG lasers. The advantage of using a DQW structure is the reduced far field angle from 41° to 24°, which enables easy beam shaping in different applications, particularly in the targeted miniaturized MOPA systems. On the other hand, a reduced SMSR can be expected from the DQW structure, due to its lower SG mode amplitudes. In addition, the DQW structure provide higher (up to 20 mW) output powers than the SQW. In regards to the grating order, a 1st order is preferred over a 2nd order grating as it provide higher output powers as well as coupling coefficients.

7.4 SG-DBR laser characterization

In the first three sections of this chapter, the SG-DBR laser was designed, its vertical structure was developed and characterizations of the developed SGs were performed. Next, two SG-DBR lasers are characterized that are based on the SQW and DQW vertical structure. Both devices were processed with 1st order gratings as these provided higher coupling coefficients and output powers, see Table 7.2 and Fig. 7.18, respectively.

Fig. 7.22 shows a photograph of the 4 mm long developed device, consisting of a 900 μm long gain section, a 100 μm long phase section, a $L_{gain} = 450$ μm long front SG, a $L_{front} = 1150$ μm long back SG, and two 750 μm long tilted RWs at each end of the laser. The tilted waveguide sections suppresses reflections to a power reflectivity of about $< 10^{-3}$. In addition, the facets of the laser are AR coated to minimize (even further) Fabry–Pérot modes and to enhance the effect of the developed SGs. The corresponding reflectivity curves can be found in Appendix H. Finally, three micro-heaters were implemented on top of the back SG, the front SG and on top of the phase section.

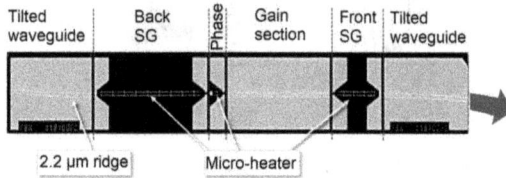

Figure 7.22: *Photograph of the developed SG-DBR laser.*

7.4.1 Electro-optical characteristics

The PUI characteristics of the two SG-DBR lasers at a heat sink temperature of $T = 25°C$ are shown in Fig. 7.23. Each laser was measured for $I_{inj} = [0, 200]$ mA with a step size of $\Delta I_{inj} = 1$ mA. The SQW based laser has a threshold current of $I_{th,SQW} = 25$ mA, a slope efficiency $S_{SQW} = 0.29$ W/A and a maximum output power of $P_{max,SQW} = 48$ mW. The DQW based laser has a slightly higher threshold current of $I_{th,DQW} = 33$ mA, a slope efficiency $S_{DQW} = 0.43$ W/A and a maximum output power of $P_{max,DQW} = 71$ mW. In both cases, the slope efficiencies were obtained through linear fits between $I_{inj} = [75, 150]$ mA, a region above threshold and below the role-over effect.

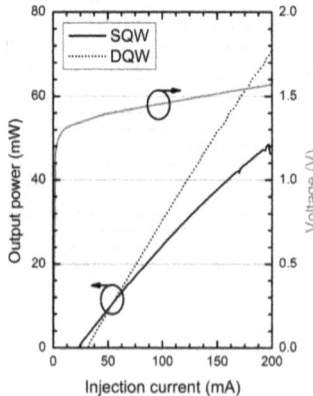

Figure 7.23: *PUI-curves of the SQW and DQW based SG-DBR lasers at $T = 25°C$.*

7.4.2 Single heater tuning

The simplest tuning scheme of a SG-laser is obtained by operating one of the two SGs which provides a discrete wavelength tuning. This is shown in Fig. 7.24 for different BM and FM heater currents. A step-like behaviour is obtained, with increasing wavelengths when the BM is tuned and decreasing when the FM is tuned, in accordance with the results of Sec. 7.1.4. In both cases, a tuning of range of 21 nm can be obtained, which is close to the predicted RMS value of 23.8 nm. The discrete steps are of the size 2.4 nm and 2.7 nm for the BM / FM tuning, respectively. These values are close to the designed values of $\Delta\lambda_{s,back} = 2.4$ nm and $\Delta\lambda_{s,front} = 2.6$ nm, see Sec. 7.1. Over this tuning range, a power variation of about 10% was measured.

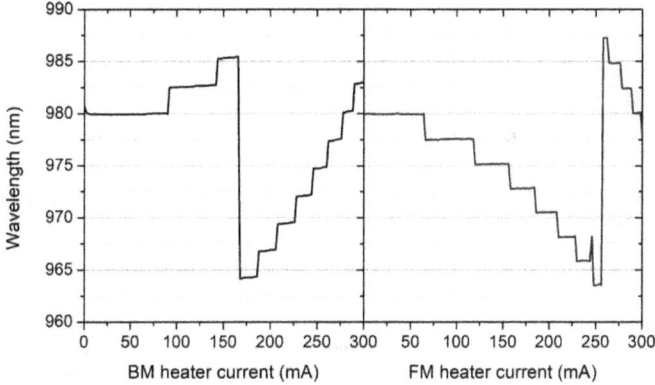

Figure 7.24: *Measured emission wavelengths at different BM (left) and FM (right) heater currents at $I_{inj} = 180$ mA and $T = 25°C$.*

Next, the tuning characteristics of the two types of vertical structures are compared with each other. Fig. 7.25 shows the emission wavelengths at selected front heater currents for the SQW (left) and the DQW (right) based lasers. Both lasers provide a tuning of about 21 nm, with a narrow emission width below 20 pm (resolution limited).

The SQW structure provides higher SMSR of about 55 dB in comparison to the 45 dB SMSR of the DQW structure. This is expected as the SQW structure have a higher coupling coefficient and should therefore have higher reflectivity, see Table. 7.2. Fig. 7.25 also shows that the DQW structure emits at longer wavelengths than the SQW structure, as was observed in Sec. 7.3.4.

This investigation indicates that SQW and DQW structures performs differently. The power characteristics in Fig. 7.23 show that the DQW structure provides higher output powers in comparison with the SQW. This result was also obtained in the power curves of the test devices with uniform DBR gratings, see Fig. 7.18. The optical performance is also different for the two structures. Although a similar tuning range of about 21 nm was obtained, the SQW structure provides higher SMSR than the DQW. Finally, the far field angle is also different for the two structures, with the SQW structure having a 41° angle while the DQW have 24°, see Sec. 7.3.2.

The overall comparison suggest that SQW based SG-lasers should be used in applications where high SMSR is required. In the targeted MOPA applications, both structures provide comparable tuning range with enough output powers to saturate a PA. The crucial point is however the far field angle, where the wide angle of the SQW structure limits the optical shaping, especially in a miniaturized MOPA system. Therefore, only the DQW structure will be considered in the following.

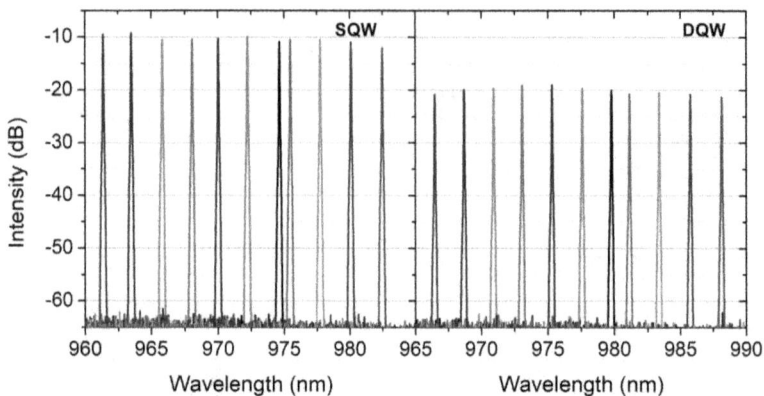

Figure 7.25: *Measured emission wavelengths at different BM heater currents of a SQW and DQW structure at $I_{\mathrm{inj}} = 180$ mA and $T = 25°$C.*

7.4.3 Dual heater tuning

Quasi-continuous wavelength tuning is obtained when both the front and the back heaters of the SG laser are operated. Thereby, a tuning map is obtained as function of the front/back heater currents, see Fig. 7.26(a). In total, quasi-continuous tuning from 964.5 nm up to 988 nm was obtained, corresponding to a tuning range of 23.5 nm. This tuning range is close to the designed $RMS = 23.8$ nm in Sec. 7.1. Over this tuning range, a mode spacing of $\delta\lambda = 115$ pm was estimated, corresponding to an effective cavity length of about 1035 μm. This value is close to the 1000 μm long manufactured cavity length (gain + phase sections) plus a small penetration depth into the two SGs. This being said, the $\delta\lambda = 115$ pm value is larger than the 90 pm calculated mode spacing of Sec. 7.1.2, where the penetration depths were estimated.

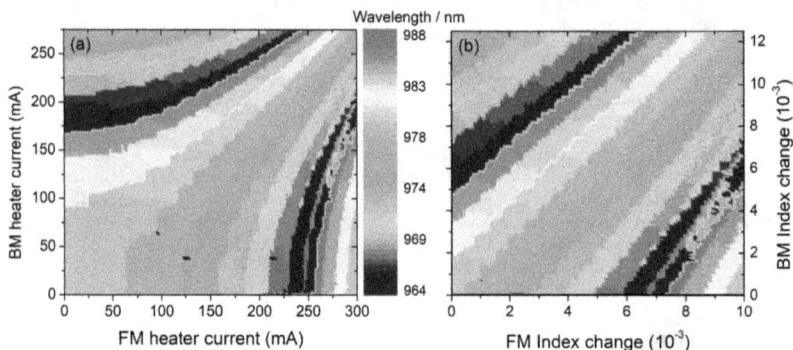

Figure 7.26: *Experimentally obtained tuning map as function of (a) different FM/BM heater currents and (b) as function of the index of refraction, at $I_{\mathrm{inj}} = 180$ mA and $T = 25°C$.*

The results in Fig. 7.26(a) show curved wavelength regions in comparison to the simulated linear regions in Fig. 7.8. This is once again due to the Joule effect of the applied micro-heaters, where the injected heater power scales quadratically with the heater current. Fig. 7.26(b) shows the tuning map as function of the index change of the FM and BM, respectively. The conversion from heater current to index of refraction was based on an estimated temperature dispersion coefficient of $\Delta n/\Delta T \approx 3 \times 10^{-4}$ 1/K [92], and a temperature gradient of $\Delta T/\Delta\lambda \approx 1/0.065$ K/nm. A good agreement is observed between the experimentally obtained results in Fig. 7.26 and the simulated in Fig. 7.8, indicating similar wavelength range coverage and wavelength regions.

7.4.4 Phase control

In addition to the two SG sections, a third passive section was implemented on the SG laser, namely the phase section. This is considered in Fig. 7.27(a) showing the emission wavelength as function of the phase heater current. The laser emits first at 978.1 nm, but between $50-125$ mA phase heater currents, emission occurs at the next SG mode (975.8 nm), and later jumps back to 978.1 nm. In addition, a reduced SMSR is seen around 250 mA where both modes are observed.

The corresponding SMSR as function of the heater current is seen in Fig. 7.27(b). The shift to the next SG mode is seen as a reduction in the SMSR from 40 to 0 dB. Likewise, a reduction in the SMSR is seen around 250 mA which goes from 50 to 15 dB, corresponding to the scenario where both modes were observed, see Fig. 7.27(a). In between, local minima/maxima are observed providing increased/decreased SMSR. This behaviour was predicted in Sec. 7.1.6 suggesting optimum and pessimum phase section indices. The improved SMSR corresponds to the scenario where a single longitudinal mode fits within the reflectivity curves. The scenario with 0 dB SMSR corresponds to the transition between two SG modes. At a heater current of 150 mA, an optimum operation setting can be found which provides the largest SMSR of 52 dB.

This being said, the simulated phase section dependency did not fully agree with the measured one due to the thermal tuning method used. The results in Fig. 7.27(a) show a wavelength change which was not predicted in the phase section simulations. This inconsistency is explained by the method of tuning. While the simulations simply consider a change in the refractive of index, the micro-heater (in addition) heats the surroundings including the reflectivity of the neighbouring SG. This leads to the wavelength changed observed in Fig. 7.27(a).

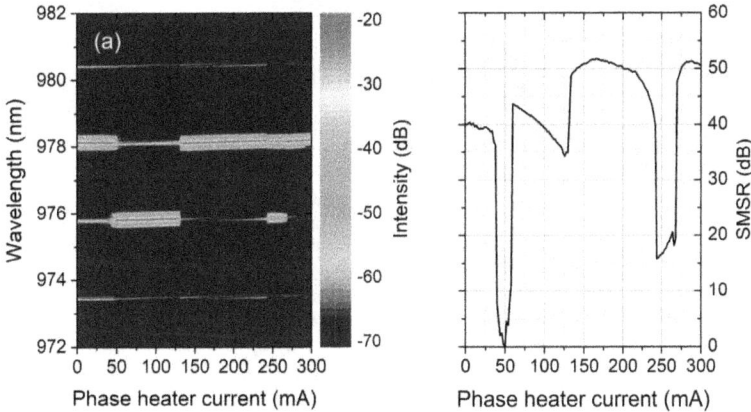

Figure 7.27: *Phase heater characteristics showing (a) the emission spectrum and (b) the corresponding SMSR of the emission wavelength at $I_{\mathrm{inj}} = 180$ mA and $T = 25°C$.*

7.5 Conclusion

In this chapter, the theory behind SG lasers was described and numerical tools to design a SG laser were presented. These include passive simulations of the SG structure using the Transfer matrix model, as well as active cavity simulations based on the coupled mode equations. These simulations describe the SG reflectivity spectra, gain, mode competition and wavelength tuning during lasing operation. Based on these simulations, a SG laser was designed to operate in single mode and have up to 23.8 nm of wavelength tuning.

Vertical structures were then grown which could provide high coupling coefficients suitable for SG laser processing. These structures were made in a two step MOVPE process, where the first step develop up to the active region and a small part of the upper waveguide layer. Parts of the active regions were then etched away in order to obtain passive regions without gain, for the processing of the phase section and the two SG structures. Here, the placement of the grating layer near the active medium was done to ensure a high coupling coefficient. Between the two processes, oxygen contamination was removed by means of in-situ etching. This was done in order to minimize losses and carrier recombination effects. The second growth process builds further on, and provides the upper waveguide, cladding and contact layers.

Different vertical and gratings structures were developed in order to understand their physical performances. The first structure was based on a SQW gain medium, which provided a large vertical far field angle of 41°. With a 1st order grating structure, a coupling coefficient of $\kappa = 500$ cm^{-1} could be obtained, while a 2nd order grating provided a coupling coefficient of about $\kappa = 300$ cm^{-1}. Although a large coupling coefficient is highly desirable in SG lasers, the large far field angle puts a challenge on the light collection. This fact makes this type of laser unsuitable for miniaturized MOPA configurations.

A second vertical structure was therefore developed utilizing a DQW structure with a larger n-side cladding layer. This structure provided a reduced far field angle of 24° which is more suitable in MOPA configurations. This however comes at the cost of a reduced coupling coefficients of $\kappa = 250$ cm^{-1} and $\kappa = 150$ cm^{-1} for the 1st and the 2nd order gratings, respectively.

Based on the design parameters and the developed vertical structures, SG lasers were manufactured on SQW and DQW based vertical structures. Both lasers were manufactured with a 1st order grating, as it provided higher output powers and coupling coefficients in comparison to the 2nd order grating.

The processed SG lasers emitted 50 and 70 mW of output powers from the SQW and DQW structures, respectively. These devices were thermally tuned using micro-heaters implemented on top of each SG. Up to 21 nm of discrete wavelength tuning could be obtained from both devices while maintaining a spectral linewidth below 20 pm. The SQW structure provided a SMSR of 55 dB while the DQW provided 45 dB. This difference is explained by the different coupling coefficients. A higher coupling coefficient provides higher reflectivity and vice versa for a smaller coefficient. Between the two lasers, the DQW was chosen as its narrow far field angle is more suitable for MOPA applications.

By operating both SG heaters, 23.5 nm of quasi-continuous tuning could be obtained with a mode spacing of about $\delta\lambda = 115$ pm. This is in agreement with the designed grating structures. By operating the micro-heater on top of the phase section, the longitudinal mode spacing can be adjusted. This results in an improved SMSR (or worsen) which once again agrees with the numerical simulations.

The demonstrated GaAs based SG lasers serve as the first and so far only demonstration of GaAs based SG lasers, in comparison to the well established InP SG lasers. The challenges of of using GaAs includes manufacturing of shorter grating periods, difficult regrowth and surface oxidation.

Chapter 8

High power tunable MOPA lasers

The described light sources so far provide low output powers (< 0.5 W) and wavelength tuning by different means. In order to obtain output powers in the watt range as required in upconversion applications, MOPA systems are constructed. This chapter describes the applied MOPA configuration, and results of three developed MOPA systems will be presented. These systems utilizes TPAs, but with different tunable MO light sources: a DBR-RW, a Y-branch DBR-RW and a SG-DBR laser.

This chapter begins with a description of the applied TPAs, including their vertical and lateral structures, mounting and facet coating. This is followed by a description of the MOPA systems, including numerical simulations of the light coupling between the MO and PA. Finally, experimental results of the three developed MOPA systems are presented and discussed.

8.1 Tapered power amplifier

Tapered diode lasers provide high output powers with nearly diffraction beams. Such devices utilize a RW section to define a narrow beam waist, while the following tapered section ensures amplification up to ten watts of output power [100]. The performance of tapered diode lasers lies between RW and BA lasers. Tapered lasers provide nearly diffraction limited beams ($M^2_{1/e^2} < 2$) with high output powers ($10 - 15$ W) [101]. RW lasers have excellent beam quality, and can emit up to 3 W of output power [42]. BA lasers on the other hand provide tens of watts of output powers at the cost of poor beam quality.

Tapered lasers can have a narrow emission linewidth when combined with e.g. a DBR grating. Alternatively, by applying an AR coating to both facets, such structures can be utilized as optical amplifiers, providing high gain and nearly diffraction limited emission.

8.1.1 Vertical structure

The vertical structure of the TPAs is based on an asymmetric super large optical cavity (ASLOC) structure, see Fig. 8.1.

The different layers were grown by MOPVE on a n-GaAs substrate, which utilize a DQW structure for increased gain amplitude. 8 nm thick InGaAs QWs were used as an active medium, sandwiched between GaAsP spacer layers. This structure emits TE-polarized light, due to the compressive strain in the DQW structure. In order to reduce the vertical far field angle, an ASLOC design was used for the confinement layers, in a similar fashion to the DQW SG laser, see Sec. 7.3. With this asymmetry, higher order lasing modes are suppressed. In addition the series resistance is reduced, which helps decrease the facet load. This structure provides a narrow vertical far field of about 14.3° at FWHM, which allows easy coupling into fibers or non-linear crystals. The temperature gradient of these TPAs was experimentally determined to be $\Delta\lambda/\Delta T = 0.3$ nm/K. This value is set by the material property, whereas the small value of $\Delta\lambda/\Delta T = 0.065$ nm/K value of the MOs is mainly set by the grating.

The figures of merit of this structure were obtained in a similar fashion to those in Sec. 3.1.1 and are summarized in Table 8.1. A large internal quantum efficiency η_i together with low internal losses α_i makes this structure suitable for amplification.

160 nm — p-GaAs Contact layer
330 nm — p-Al$_{0.85}$Ga$_{0.15}$As Cladding layer
800 nm — p-Al$_{0.35}$Ga$_{0.65}$As Waveguide layer

InGaP Spacer layer — 5 nm
In$_{0.16}$Ga$_{0.84}$As QW — 8 nm
Active region — InGaP Spacer layer — 7.5 nm
In$_{0.16}$Ga$_{0.84}$As QW — 8 nm
InGaP Spacer layer — 5 nm

4850 nm — n-Al$_{0.35}$Ga$_{0.65}$As Waveguide layer
500 nm — n-Al$_{0.45}$Ga$_{0.55}$As Cladding layer
300 nm — n-GaAs buffer
n-GaAs Substrate

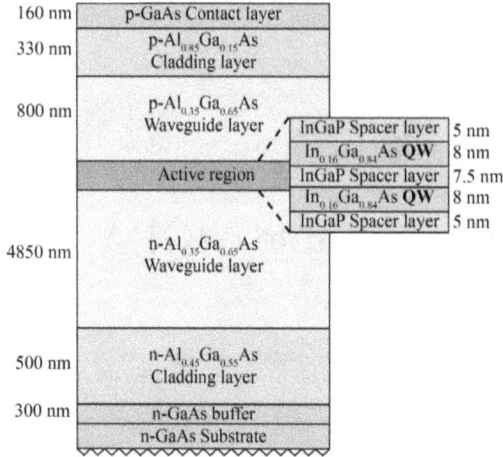

Figure 8.1: *Vertical structure of the TPA.*

Table 8.1: *Material data parameters of the TPA vertical structure.*

Parameter	Symbol	Value
Internal quantum efficiency	η_i	≈ 0.91
Internal losses	α_i	0.3 cm^{-1}
Transparency current density	J_{TR}	134 A/cm^2
Modal gain coefficient	Γg_0	23 cm^{-1}
Characteristic temperature of the threshold current	T_0	160 K

8.1.2 Lateral structure

The applied TPA devices of this chapter are 6 mm long and consists of two sections; a 5 μm wide, 2 mm long RW section, followed by a 4 mm long TPA section with a tapered angle of 6°, forming a 422 μm wide aperture (front facet). The chosen taper angle corresponds to the diffraction angle of the light exiting the RW section, and thus provide a smooth transition between the two sections. Fig. 8.2 shows a photograph of a TPA, indicating the two separated gold contacts of the RW and TPA sections. This allows for individual contact and control of the current injection, which can provide improved beam quality at certain current settings [102].

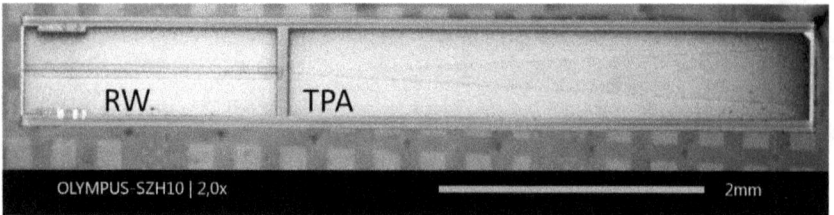

RW. TPA

OLYMPUS SZH10 | 2,0x   2mm

Figure 8.2: *Photograph of the TPA showing the lateral structure.*

8.1.3 Mounting and facet coating

Similarly to the previously described lasers, the TPA chips are mounted p-side up on CuW sub-mounts, a material chosen due to its comparable expansion coefficients with the GaAs chip material. In addition, the TPAs have another CuW top-mount (sandwiched), which allows a more uniform current injection into the relatively wide TPA section. Fig. 8.3 shows the mounting of these devices, including the individual bonding wires, which ensures a uniform current injection into the RW section, and into the top-mount of the TPA section.

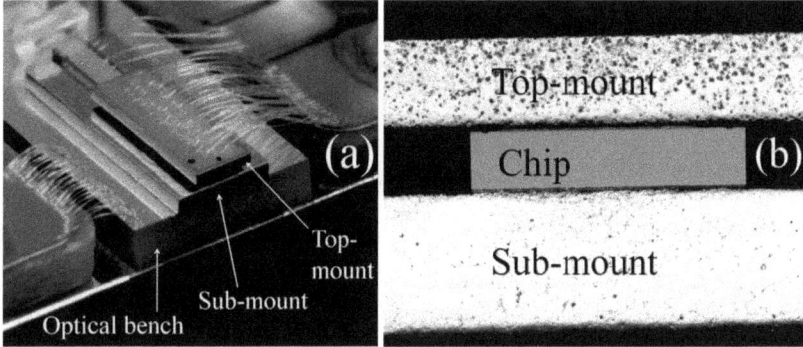

Figure 8.3: *Photographs of (a) the mounting and wire bonding and (b) and the sandwich mounting and facet of the TPA device.*

Both facets of the TPAs were cleaned by applying atomic hydrogen and a sealing process with ZnSe before applying the AR coating [67]. This coating was made to ensure pure amplification and to minimize lasing from the PA itself.

8.1.4 Gain spectrum

The vertical structure of the TPAs was chosen due to its gain shape which provide between 20 and 25 nm of gain bandwidth around 976 nm. The ASE of such an amplifier is shown in Fig. 8.4, measured at different TPA currents for $I_{RW} = 200$ mA at $T = 20°$C. The shown gain bandwidth fits well with the achieved tuning ranges of the different lasers of this work. This being said, for a wider tuning range a SQW or a varying MQW structure would be more suitable as it provide a wider gain bandwidth.

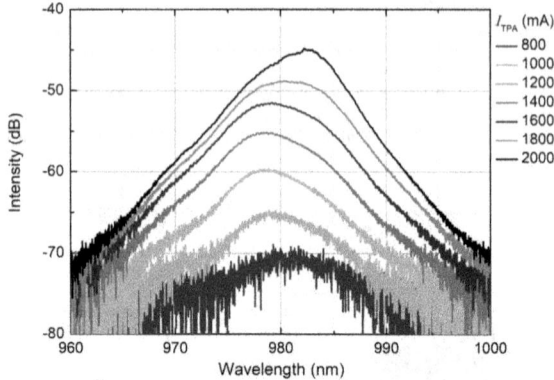

Figure 8.4: *Measured gain spectra at different I_{TPA} currents, measured with $I_{RW} = 200$ mA and at $T = 20°C$.*

8.2 Hybrid MOPA configuration

A hybrid MOPA configuration was applied between the tunable MO and the amplifying TPA. In general, one key advantage of using a hybrid approach, instead of a monolithic, is the capability of optimizing the epitaxial structures of both components independently. In regards to tunable MOPA configurations, the motivation behind a hybrid approach is to minimize the thermal decoupling between the two components. As the PA is injected with a relatively high current (multiple of amperes), spacing the MO and PA apart reduces the thermal feedback. On the other hand, the micro-heaters of the MO provide a relatively high temperature gradient (see chapter 6), which in a monolithic MOPA configuration would result in a reduced output power during the wavelength tuning. In addition to thermal feedback, by separating the two components, the optical feedback between the MO and PA can be reduced.

The emitted light by the MO was coupled into the PA using a system of optical lenses, chosen according to optical simulations performed with the software *WinABCD* by BeamXpert [103]. As the name suggests, this software uses a ray-transfer (ABCD) matrix method, to describe the change of an optical beam after propagating through a set of lenses and/or other optical components [104].

The MO is defined in this software by its aperture, which is set by the vertical structure and its RW width, and is collimated by a set of fast/slow axis collimation (FAC/SAC) cylindrical micro-lenses. Among the investigated lenses, an FAC lens with the dimensions $L \times W \times H = 1.5$ mm $\times 0.8$ mm $\times 1$ mm, a focal length of $f_{FAC} = 0.6$ mm and a numerical aperture (NA) of 0.8 was chosen. The chosen SAC lens have the dimensions $L \times W \times H = 1.5$ mm $\times 1$ mm $\times 1.5$ mm, a focal length of $f_{SAC} = 2.3$ mm and a NA of 0.3. The positioning of the lenses was done using a *SmarPod 6D* positioner from SmarAct, which has a resolution of about 10 nm [105].

The PA is defined as a slit by its aperture, and the coupling of the collimated light of the MO was estimated by the transmission through this slit, see Fig. 8.5. This transmission can be estimated both as the total transmitted power, or as the transmitted power which resembles a Gaussian profile inside the slit. A similar set of lenses (SAC/FAC) were chosen, which provides a transmission/coupling of 97.5% along the horizontal axis and a 93.1% transmission along the vertical axis. According to the Gaussian profiled transmission, a transmission of about 88.7% was obtained for both axes. This coupling corresponds to a 1:1 imaging of a MO aperture (spot) that is 2.5 μm wide and 2 μm tall, into the PA aperture of 5 μm wide and 2 μm tall. The vertical apertures of the MO and PA fits with one another, and the FACs makes a 1:1 imaging into the PA. The wider RW of the PA makes it easy to couple the MO light, and offers spatial filtering as the divergence of the MO occurs at the start of the RW section of the PA.

While the slit representation of the PA is not complete, it gives a good estimate of the amount of coupling. A similar concept has been used to design a miniturized second harmonic generation (SHG) laser system using cylindrical micro-lenses [106].

82

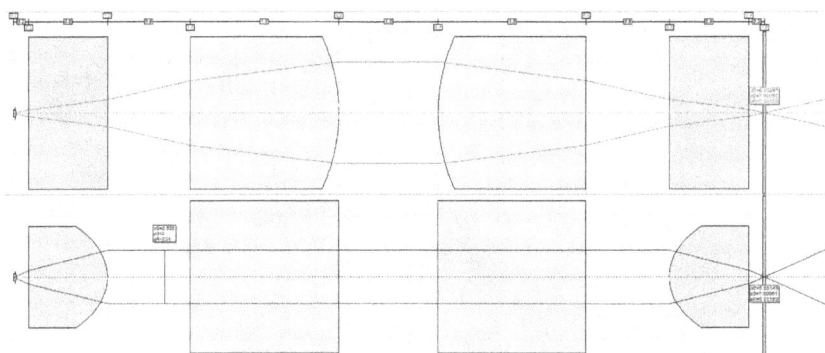

Figure 8.5: *Optical simulation of the optical coupling between the MO (left) and PA (right), using the program WinABCD.*

As mentioned earlier, the MO and PA were mounted p-side up on CuW sub-mounts during characterization and in the MOPA configuration. These chips were then mounted on top of an AlN heat spreader mount using AuSn solder. This heat spreader serves as the base for the optical bench, and has the dimensions $L \times W \times H = 20$ mm $\times$ 5 mm $\times$ 1 mm. Likewise, the collimation and focusing lenses were glued on top of the optical bench using UV-harden glue.

The configuration and the dimensions of the MOPA system are shown in Fig. 8.6. The disadvantage of this arrangement is the necessary high-precession alignment of both the semiconductor components and of the micro lenses. During the hardening process, the glue shrink a couple of micrometers depending on the amount of applied glue, as well as the power and exposure time of the UV light. Therefore, this shrinkage must be accounted for, especially in the case of the FAC lenses where the vertical position is crucial. Typical accuracies between 300 to 500 nm were achieved for the different MOPA systems of this work.

Photographs of the developed MOPA systems can be found in Appendix E.

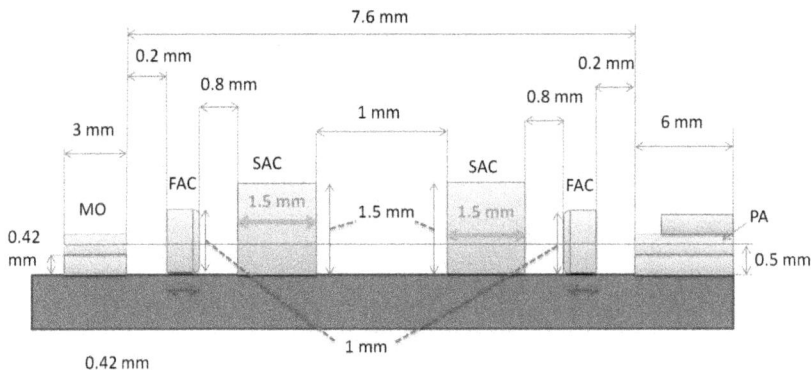

Figure 8.6: *Sketch of the MOPA system in a micro-bench setup.*

8.3 DBR-RW based MOPA system

The first developed MOPA system is based on a tunable DBR-RW laser (similar to the reference laser of Sec. 3.4) combined with a TPA, see Fig. 8.7. This section describes the performance of this system including its power, spectral, spatial, tuning and stability performances.

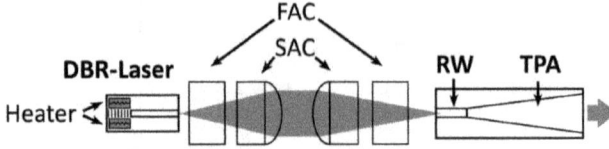

Figure 8.7: *Sketch of DBR-based MOPA system.*

8.3.1 Electro-optical characteristics

The MO used in this system is similar to the reference laser, however its front facet is coated with 30% instead of the 5% reflectivity used in earlier chapters, see Appendix H. This configuration with higher front facet reflectivity was used, as it proved to be more stable in terms of wavelength stabilization. This is explained by the reduced coupling between the back reflected light from the PA and into the MO (and the optical elements), as it suppresses a possible cavity arising between the front facets of the PA and the MO.

The emitted output power of the MO (with the increased front facet reflectivity) is shown in Fig. 8.8, together with the output power of the MOPA system as function of the MO current/output power. From this measurement, a seed power of about $P_{MO} = 30$ mW ($I_{RW} = 100$ mA) was chosen although typical saturation power of TPAs lies around $15-20$ mW [22]. This over-saturation setting was chosen to ensure a uniform amplification from the PA, even during the wavelength tuning. As the output power of the MO drops when applying the micro-heaters (see chapter 6), the reduced output power of the MO will still be enough to saturate the PA.

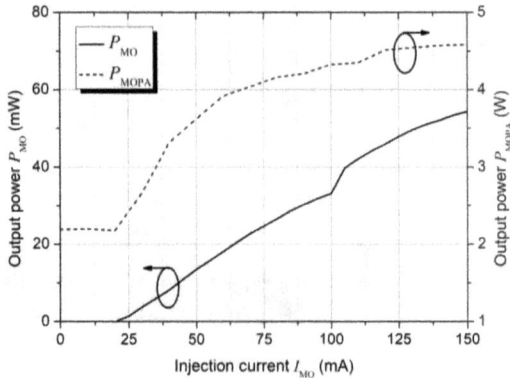

Figure 8.8: *Power characteristics (left axis) of the MO and (right axis) of the emitted output power of the MOPA system at $I_{RW} = 300$ mA, $I_{TPA} = 6.5$ A and $T = 15°C$.*

The output power of the MOPA system as function of the TPA current I_{TPA} is seen in Fig. 8.9, for a RW injection current of $I_{RW} = 300$ mA at a heat sink temperature of $T = 15°C$. This temperature was chosen as it provided the least spectral detuning between the MO and PA. In addition, this lower temperature provides higher output powers in comparison to e.g. 25°C. It is seen in Fig. 8.9 that at an amplifier current of $I_{TPA} = 8.5$ A, a maximum output power of 6.2 W was obtained. At higher TPA currents, the system becomes unstable and self-lasing from the PA was observed. This is explained by the different temperature gradients $\Delta\lambda/\Delta T$ (0.065 vs. 0.3 nm/K) of the MO and PA, respectively. I.e. the stronger red-shift of the gain of the PA, relative to the shifting of the DBR grating of the MO.

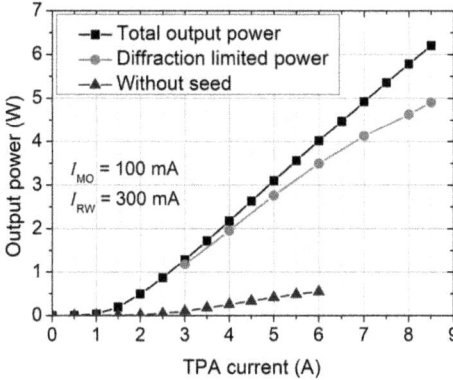

Figure 8.9: *Power characteristics of the MOPA as function of the amplifier current I_{TPA}. The black curve (squares) shows the total emitted power, whereas the red curve (circles) shows the diffraction limited power of the PA, both when operated with $I_{MO} = 100$ mA and $I_{RW} = 300$ mA at $T = 15°C$. The blue curve (triangles) shows the output power of the PA before attaching the micro-lenses and without being seeded.*

In addition, Fig. 8.9 shows the amount of power in central lobe as function of the amplifier current (red circles). These values were obtained from the lateral intensity distribution of the emitted light, as will be shown next in the next section. Finally, the output power of the TPA (blue triangles) before attaching the micro-lenses, and without being seeded is also shown in Fig. 8.9. This measurement indicates an upper limit of the amount of ASE of the PA, although it is expected to reduce when the PA is seeded.

The corresponding emission spectrum at $P_{MOPA} = 6.2$ W is shown in Fig. 8.10. This spectrum indicates an emission wavelength of 971.82 nm, with a FWHM spectral width of about 17 pm (spectrometer resolution limit).

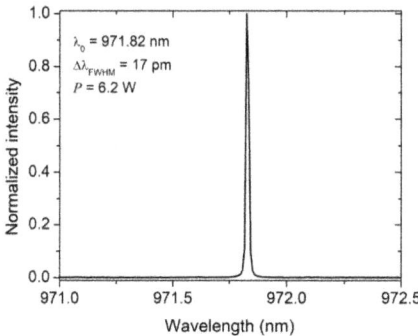

Figure 8.10: *Spectrum of the MOPA system at an output power of $P_{MOPA} = 6.2$ W at $T = 15°C$.*

8.3.2 Beam quality

The lateral beam quality of the MOPA system was characterizied according to the method of the moving slit [68]. The normalized lateral intensity distributions of the near field, beam waist and far field at $P_{MOPA} = 6.2$ W are shown side-by-side in Fig. 8.11. Note that due to the tapered structure, the near field and beam waist are at two different positions; the beam waist is close to the front of the RW section, while the near field is measured at the front facet of the PA. In addition, this astigmatism (≈ 1.5 mm) increases approximately linearly with the TPA currents by 15 μm/A.

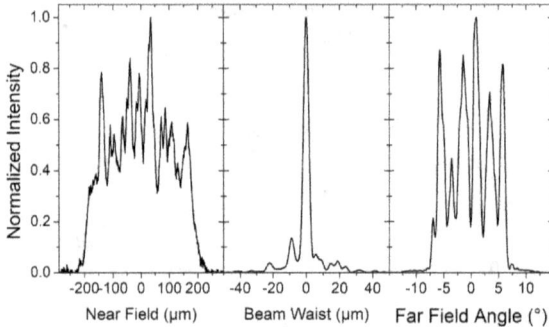

Figure 8.11: *Normalized intensity distributions of the near field (left), beam waist (center) and far field (right) at $P_{MOPA} = 6.2$ W and $T = 15°$C.*

A near field width of about 392 μm was measured at $1/e^2$, which is in good agreement with the 422 μm wide front aperture of the TPA. The measured beam waist profile shows a number of lateral side lobes and a defined central lobe with a width of about 6.4 μm ($1/e^2$). From the beam waist measurements, the power in central lobe was estimated to lie between $P_{CL} = [91, 79]\%$ for $I_{TPA} = [3.0, 8.5]$ A. The corresponding diffraction limited output powers are shown in Fig. 8.9 as red circles. The far field angle of about 13.7° together with the beam waist width provide a beam propagation factor of $M^2_{1/e^2} = 1.3$ in the slow axis at $I_{TPA} = 8.5$ A.

8.3.3 Wavelength tuning

Wavelength tuning is the next parameter considered. As described in chapter 6, resistor based micro-heaters were embedded on top of the DBR section, which can provide thermal wavelength tuning. Fig. 8.12 shows the emission wavelength of the MOPA system as function of the heater current I_{Heater}. It is seen that the wavelength of the MOPA system can be tuned from 971.82 nm up to 977.23 nm, thus covering a 5.5 nm tuning range with a mode spacing of about $\delta\lambda = 64$ pm. The 5.5 nm tuning corresponds to a temperature gradient of about 85 K as described by the $\Delta\lambda/\Delta T = 65$ pm/K (see Sec. 6.1).

A quadratic tuning behaviour is once again seen, as the induced electrical power scales quadratically with current. Fig. 8.12 shows poor SMSR at shorter wavelengths ($I_{Heater} < 150$ mA), which is improved as the wavelength of the MO increases. This is also seen in the inset of Fig. 8.12 for the $I_{Heater} = 0$ mA. This is due to a spectral mismatch between the MO and PA, which is reduced at longer MO wavelengths. This effect can be reduced by operating at lower temperatures, as $\Delta\lambda/\Delta K$ is stronger for the PA (set by the gain) than for the MO (set by the grating). However, the chosen $T = 15°$C is more suitable during applications as it can be obtained without the risk of water condensation (especially during summer) at the laser facets. Nonetheless, single mode operation was observed at most heater currents, expect in between the individual mode jumps. The $\delta\lambda = 64$ pm mode spacing corresponds to the MO effective cavity length of about 2.05 mm (as was already shown in Sec. 3.4).

The output power during the wavelength tuning is another important parameter, which is shown in Fig. 8.13 as function of the heater current. It is seen that an almost constant output power was measured,

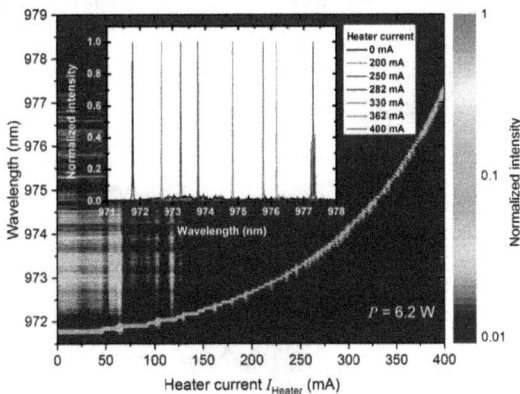

Figure 8.12: *Contour plot of the wavelength as function of the heater current at* $T = 25°C$. *Inset: emission wavelengths at selected heater currents.*

with an average power of (6.22 ± 0.01) W, maintained over the entire tuning range. This variation corresponds to a SDOM of ± 10 mW ($\pm 0.16\%$). This being said, a tendency of decreasing output power is seen at $I_{Heater} > 390$ mA, which is likely due to the redshift of the MO relative to the gain maximum of the PA. Finally, the heater voltage drop is also shown in Fig. 8.13, which at a heater current of $I_{Heater} = 400$ mA has a voltage drop of 7.1 V, resulting in an induced heater electrical power of 2.84 W and a resistance of 18 Ω.

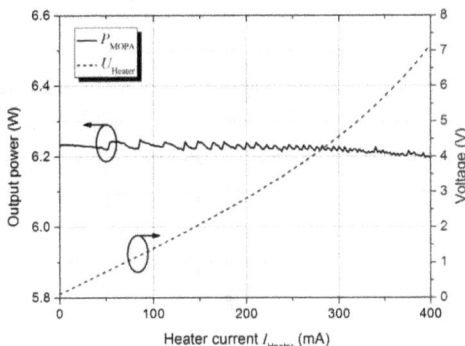

Figure 8.13: *MOPA output power (left axis) and heater voltage (right axis) as function of the heater current* I_{Heater} *at* $T = 15°C$.

8.3.4 Stability

Finally, stability is a crucial parameter for most laser applications. Therefore, the output power of the MOPA system was measured once every second for a period of 1 hour, and a mean output power of 6.240 W $\pm$ 2 mW was obtained, corresponding to a SDOM of $\pm 0.03\%$, see Fig. 8.14.

Likewise, the emission spectrum was measured over the same time period as shown in the inset of Fig. 8.14. An emission wavelength of 971.86 nm$\pm$17 pm was emitted and maintained, limited by the 17 pm spectral resolution of the spectrometer.

In this measurement, no precautions were made to minimize temperature fluctuations, and this level of stability indicates that there is no need for an an optical isolator between the MO and PA. Nonetheless, an optical isolator might still be required after the MOPA system, in e.g. non-linear frequency conversion applications, where the light is tightly focused into a non-linear crystal [18].

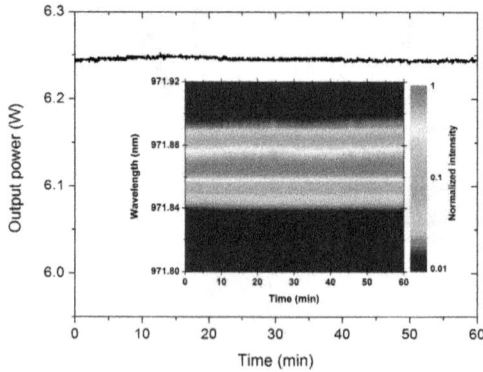

Figure 8.14: *Stability measurement of the output power for a period of 1 hour at $T = 15°C$. Inset: corresponding emission wavelength.*

8.3.5 Summary

In conclusion, the presented MOPA system which combines a tunable DBR-laser with a TPA provides an output power of 6.2 W at a wavelength of 971.82 nm, with a spectral width of about 17 pm. The emitted light is nearly diffraction limited with a propagation factor of $M_{1/e^2}^2 = 1.3$ in the slow axis, and has 4.9 W of diffraction limited power. The embedded micro-heater on top of the grating section of the MO provides up to 5.5 nm of electrically controlled wavelength tuning. Over this tuning range, an average output power of 6.22 W $\pm$ 0.01 W was maintained. The stability of the system was tested for a period of 1 hour, which emitted an average output power of 6.240 W $\pm$ 2 mW and a wavelength of 971.86 nm $\pm$ 17 pm. Thus no indications of instabilities or feedback issues were observed.

8.4 Y-branch based MOPA system

The next MOPA system was based on a Y-branch DBR-RW laser with a Sinusoidal S-bends. The investigation in chapter 4 showed that the three types of Y-branch lasers had similar performances, with the Sin having slightly better beam quality. Once again, the MO utilized has a 30% front facet reflectivity, which is more stable (than the 5% coating) against back reflections from the PA. The higher reflectivity couples less output powers, while the spectral properties shown in chapter 4 were unchanged. As motivated earlier, the advantage of using a Y-branch laser compared to a regular/standard DBR laser, is the extended combined wavelength tuning range.

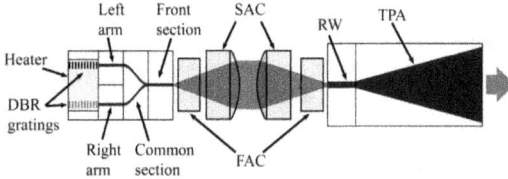

Figure 8.15: *Sketch of the Y-branch based MOPA system.*

8.4.1 Electro-optical characteristics

The investigation of the Y-branch lasers in chapter 4 showed that comparable optical characteristics from both the left and right arms were obtained, which is a big advantage in a MOPA system. In addition, it was shown that spectral single mode was obtained over the entire injection current range by not operating the Y-section of these lasers. Therefore, the MO was operated by injecting 100 mA into each arm, and 50 mA into the front section, see Fig. 8.15. This provided an output power of 24 mW and 21 mW from the left/right arm, respectively. Once again, these values are chosen to be above the saturation power ($\approx$ 15 mW), to ensure saturation even during the wavelength tuning.

The corresponding output power of the MOPA system as function of the TPA current is shown in Fig. 8.16(a) for a RW current of $I_{RW} = 300$ mA at a heat sink temperature of $T = 25°C$. As can be seen, the MOPA system emits comparable output powers regardless of being seeded from one (left or right) or even when seeded from both arms simultaneously. Simultaneous seeding was done by setting $I_{front} = 50$ mA, $I_{left,arm} = 100$ mA and $I_{right,arm} = 100$ mA.

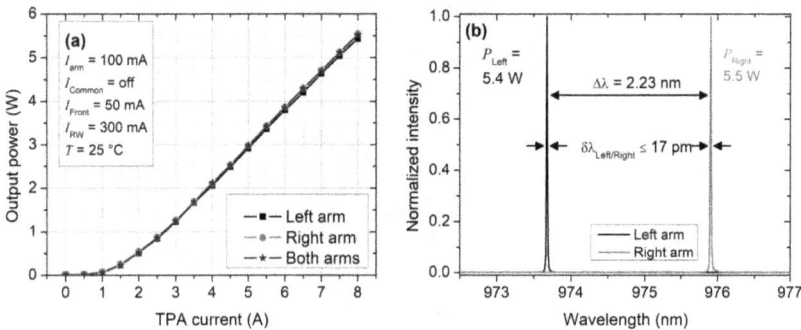

Figure 8.16: *Electro-optical characteristics of the MOPA system showing (a) the MOPA output power and (b) the corresponding emission wavelengths of the individual arms at maximum output powers.*

The maximum output power for the left, right, and both arms at a TPA current of $I_{TPA} = 8$ A were $P_{Left} = 5.4$ W, $P_{Right} = 5.5$ W and $P_{Both} = 5.5$ W, respectively. This is another indication that the PA was being saturated properly, as the case of dual arms provides about twice the seed power compared to single arm operation, however, no increase in output power was observed.

The corresponding emission spectra of the left and right arms at their corresponding maximum out powers are shown in Fig. 8.16(b). The left arm emits at $\lambda_{Left} = 973.67$ nm while the right arm emits at $\lambda_{Right} = 975.91$ nm, both having FWHM spectral widths below 17 pm (spectrometer resolution limit).

8.4.2 Beam quality

The normalized lateral intensity distributions of this MOPA system are comparable to the ones in Fig. 8.11 of the DBR-RW based MOPA system. The spatial characteristics are mainly set by the TPA and although different MO were used, comparable spatial characteristics were obtained.

In the case of the Y-branch MOPA system, a near field width of about 390 μm $(1/e^2)$ was measured, and the beam waist profile showed a number of lateral side lobes, with a 6.4 μm $(1/e^2)$ wide defined central lobe. The far field angle was about 13.7°, which provides a beam propagation factor of $M^2_{1/e^2} = 2.2$ in the slow axis. The power content within the central lobe was 72% at a TPA current of 8 A, which corresponds to a diffraction limited output power of approximately 4 W.

8.4.3 Wavelength tuning

Wavelength tuning of the MOPA system was once again obtained by operating the micro-heaters, that were implemented on top of the DBR gratings of the MO. As described in chapter 6, the widest wavelength tuning was obtained when the heaters are connected in series. At maximum output power from both arms, the heater was operated with currents between $I_{Heater} = [0, 750]$ mA, with a step size of $\Delta I_{Heater} = 5$ mA. The corresponding emission spectra (superimposed) are shown in the contour plot in Fig. 8.17.

The left arm can be tuned from $\Delta\lambda_{Left} = [973.70, 981.20]$ nm, and the right arm from $\Delta\lambda_{Right} = [975.89, 983.40]$ nm, both with a mode spacing of about $\delta\lambda = 52$ pm. This corresponds to a tuning of $\Delta\lambda = 7.5$ nm from each arm, and to a temperature gradient of about 115 K, estimated according to $\Delta\lambda/\Delta T = 65$ pm/K. Note that in this case, the micro heaters were operated up to 750 mA, whereas in chapter 6 they were only operated up to 500 mA, which explains the wider tuning ranges.

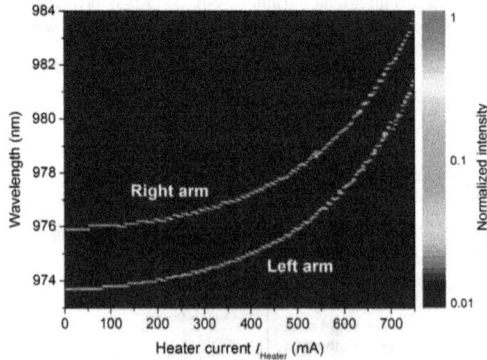

Figure 8.17: *False color contour plot of the spectral tuning of the emission wavelength of the individual arms (super imposed) as function of the heater current at $T = 25°C$.*

With the spectral distance of 2.23 nm between the two arms, and in combination with the micro-heaters, a combined wavelength tuning of $2.2 + 7.5$ nm $= 9.7$ nm was obtained from 973.70 nm (left arm + $I_{Heater} = 0$ mA) up to 983.40 nm (right arm + $I_{Heater} = 750$ mA).

Up to a heater current of approximately $I_{Heater} = 60$ mA, the change of the emission wavelength was within the resolution limit of the spectrometer, see Fig. 8.17, and a quadratic tuning behaviour is once again observed (see chapter 6). At a heater current of $I_{Heater} = 750$ mA, a voltage drop of 4 V was measured, thus inducing 3 W of electrical power into the heaters with a resistance of 5.3 Ω.

Single mode operation was observed at most heater current steps, expect in between some of the mode hops. Fig. 8.18 shows emission spectra of the left (bottom) and of the right (top), at selected heater currents. These measurements were obtained using a *Yokogawa—AQ6373* optical spectrum analyzer, with a resolution limit of 20 pm. A SMSR of about $15 - 25$ dB was observed.

The difference in SMSR between the two arms are likely due to a different coupling efficiency from the two MO arms. Compensation was made during the lens positioning to ensure comparable coupling from both MO arms. However, it is likely that the final lens position provides a better coupling from one of the two arms, translating to a better SMSR.

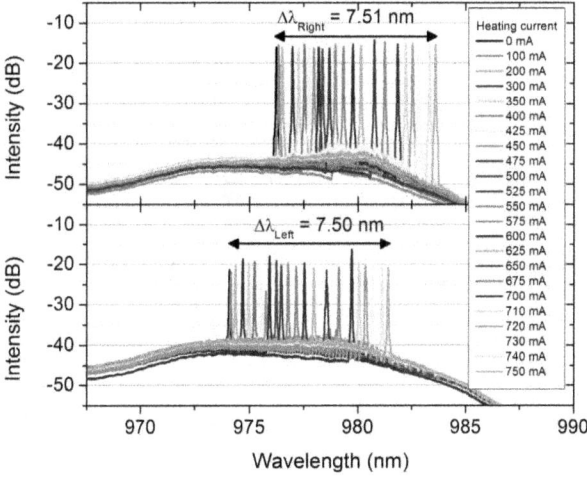

Figure 8.18: *Optical spectra of the right (top) and left (bottom) arms as function of the heater current at* $T = 25°C$.

The output power during the wavelength tuning is shown in Fig. 8.19. The left arm emits an average output power of $P_{Left} = 5.41$ W ± 0.03 W with a SDOM power variation of about $\pm 0.5\%$. The right arm emits an average power of $P_{Left} = 5.48$ W ± 0.01 W, with a SDOM of $\pm 0.2\%$.

As mentioned earlier, the MOPA system can also be operated as a dual-wavelength source when being seeded from both arms simultaneously. Fig. 8.20 shows the emission spectra of the dual-wavelength MOPA system at three different heater currents. A SMSR of about $15 - 25$ dB was observed, while no additional lasing peaks were seen in between the two wavelengths.

8.4.4 Stability

Finally, the stability of the system was investigated by measuring the output power once every second for a duration of 1 hour. Fig. 8.21 shows the normalized output power once every minute, for clarity of the stability measurement. A power SDOM of 0.03% and 0.02% were estimated for the left and right arms, respectively. Under dual wavelength operation, a power variation of 0.09% was observed over 1 hour. This increased SDOM is likely due to an additative power (in)stability effect between the two arms.

Over the same period, the emission wavelengths were measured to be $\lambda_{Left} = 973.71$ nm and $\lambda_{Right} = 975.99$ nm, both maintained within the 17 pm resolution limit. As in the case of the DBR-RW based

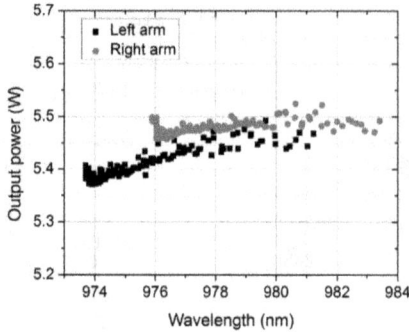

Figure 8.19: *Output power versus the emission wavelength for the two arms under individual arm operation at* $T = 25°C$.

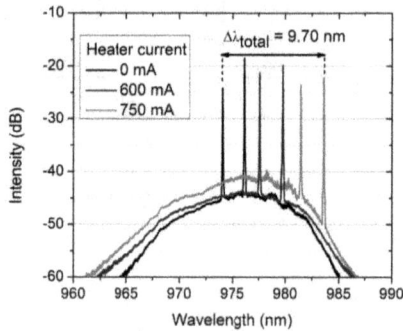

Figure 8.20: *Tuning spectra during dual-wavelength operation at* $T = 25°C$.

MOPA system, no precautions were taken to minimize temperature fluctuations or to stabilize the system during this stability measurement.

8.4.5 Summary

A compact high power MOPA system was presented which utilizes a Y-branch DBR-RW laser as MO and a TPA. It emits at two wavelengths; $\lambda_{Left} = 973.67$ nm and $\lambda_{Right} = 975.91$ nm, with a spectral linewidth limited by a 17 pm resolution limit. The emitted light of the MO was collimated and coupled into the PA using a set of cylindrical micro-lenses, constructed onto a 25 mm × 25 mm CCP mount. The MOPA system emits about 5.5 W of output power, with a beam propagation factor of $M^2_{1/e^2} = 2.2$ in the slow axis, resulting in a power in the central lobe of about 4 W (72%).

By applying the micro-heaters, each arm can be tuned by 7.5 nm, and up to 9.7 nm of combined quasi-continuous wavelength tuning can be achieved from the two arms (spaced by 2.23 nm apart). Over this tuning range, an almost constant output power was maintained, with a power variation of only 0.5% and 0.2% from the left and right arms, respectively.

In addition, the system can operate as a dual-wavelength laser when seeded from both arms simultaneously, without showing any sign of wavelength fluctuations or self-lasing effects from the PA.

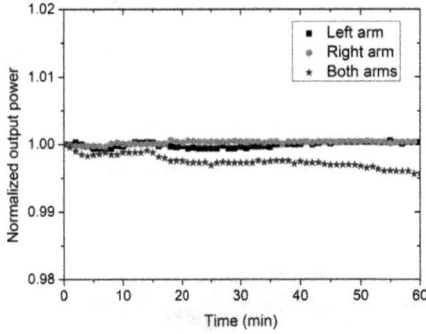

Figure 8.21: *Output power measured for a period of 1* h *at* $T = 25°C$.

Finally, the stability of the system was tested for a period of 1 hour, and very low power fluctuations of the order of 0.03% and 0.02% were observed for the left and right arms, respectively. Over the same time period, no wavelength fluctuations were observed within the 17 pm resolution limit.

8.5 Sampled-grating based MOPA system

A SG based MOPA system was also developed in this work. The SG used in this MOPA system has a DQW as active medium with a 1st order grating, which provides a reduced far field angle of 28°, see Sec. 7.3.

The scheme of the SG-MOPA is similar to the previous systems and is shown in Fig. 8.22. As described earlier, the MO has a 2.2 μm wide RW and consists of four longitudinal sections: a 900 μm long gain section, a 100 μm long phase section, a 1000 μm long back SG and a 450 μm long front SG.

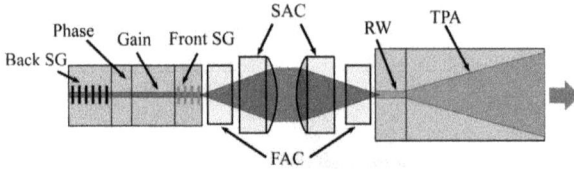

Figure 8.22: *Sketch of the SG-based MOPA system.*

8.5.1 Electro-optical characteristics

The PA was seeded with a measured MO output power of 25 mW, and the corresponding output power of the MOPA as function of the TPA current is shown in Fig. 8.23. A total output power of $P_{MOPA} = 1040$ mW was obtained at amplifier currents of $I_{TPA} = 6$ A and $I_{RW} = 5$ mA, when operated at a heat sink temperature of $T = 15°C$. These low output power values are set by the low I_{RW} current value.

At higher amplifier currents, in particular through the RW section, self-lasing from the PA was observed. This is likely due to a cavity arising between the PA front facet and the front SG of the MO, and this effect (of self-lasing) will be explained later in this section.

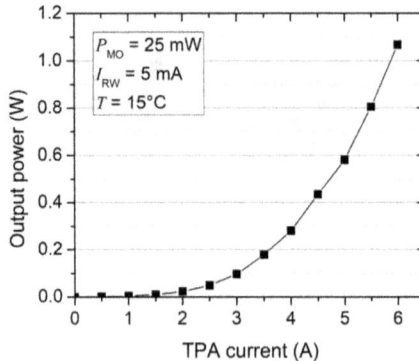

Figure 8.23: *Power characteristics of the MOPA system as function of the amplifier current.*

8.5.2 Beam quality

The spatial characteristics of the emitted light were characterized once again according to the moving slit method [68]. The near field has a width (measured at $1/e^2$) of 425 μm, in agreement with the TPA front aperture. The beam waist has a width of about 7.4 μm, containing 74% of the total output power, and a

15.5° far field angle was measured. This results in a nearly diffraction limited beam with a propagation factor of $M_{1/e^2}^2 = 1.6$ in the slow axis.

8.5.3 Single heater tuning

The emission spectrum at $P_{MOPA} = 1040$ mW is denoted as (1) in Fig. 8.24. The MOPA system emits at a wavelength of $\lambda = 979.08$ nm with a FWHM spectral width smaller than 20 pm (resolution limit).

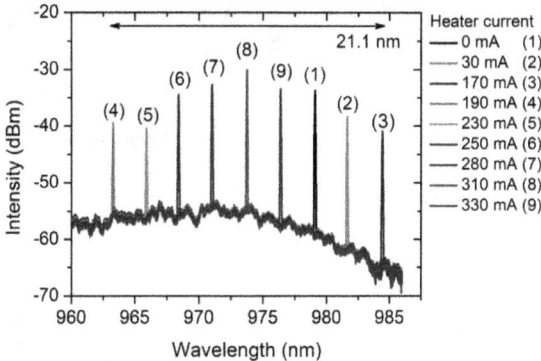

Figure 8.24: *Emission spectra during single (back) heater tuning, with $I_{TPA} = 6$ A, $I_{RW} = 5$ mA and $T = 15°C$.*

By operating one of the two micro-heaters embedded on top of each SG, a discrete wavelength tuning of $\Delta\lambda = 21.1$ nm with a peak-peak distance of about 2.3 nm can be obtained, see Fig. 8.24. The laser emits first at $\lambda = 979.08$ nm (1) which then jumps to the next SG mode (2), (3), ... , (9). By further increasing the heater current, the laser emits at wavelength (1) once again. Note that over this tuning range, a spectral linewidth below 20 pm was maintained.

The lower output power obtained from the SG-MOPA system (relative to the other two MOPA systems), is due to the low injection currents of the PA. As mentioned earlier, self-lasing effects took place at increasing injection currents, and this is shown in Fig. 8.25 for the cases of $I_{RW} = 50$ mA and at 100 mA. As can be seen, additional emission peaks were observed for both cases, lying in between the SG modes. Nevertheless, this effect is reduced with decreasing injection current and the $I_{RW} = 5$ mA setting was chosen as a compromise, see Fig. 8.24. Note that the additional emission peaks were mainly present in the case of the (4) - (6), SG-modes, i.e. at the shortest emission wavelengths.

One of the differences between this MOPA system and the previous two is the front facet reflectivity. While the two other systems had a 30% front facet reflectivity, this MO has a SG and an AR coating. The front SG had a maximum reflectivity of about 60% (see Sec. 7.1). This high reflectivity is likely the cause for the arising cavity between the MO and PA. In addition, an improper seeding is expected when injected a low current into the RW section of the PA.

By reducing the I_{RW} injection current to this level, the light of the MO undergoes a section with absorption. This reduces additional lasing peaks due to a cavity arising between the front facets of the MO and PA, but comes at the cost of overall reduced power amplification.

8.5.4 Dual heater tuning

By simultaneous heating of both SGs, a tuning map is obtained as function of the individual heater currents, see Fig. 8.26(a). In this measurement, the heater currents were varied between $I_{heater} = [0, 350]$ mA with a step size of $\Delta I_{Heater} = 5$ mA.

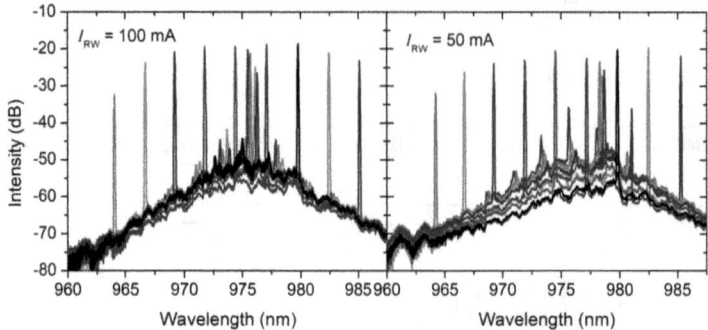

Figure 8.25: *Emission spectra during single (back) heater tuning, with $I_{RW} = 100$ mA (left) and $I_{RW} = 50$ mA (right) both with $I_{TPA} = 6$ A at $T = 15°C$.*

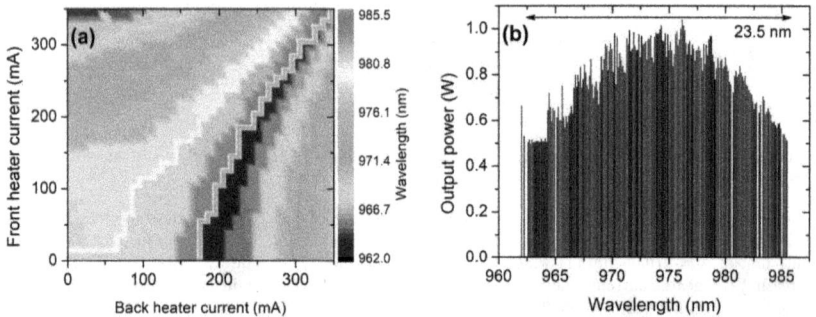

Figure 8.26: *(a) Tuning map as function of the front/back heater currents and (b) the corresponding output power at the individual emitted wavelengths at $T = 15°C$.*

In total, $\Delta\lambda = 23.5$ nm of quasi-continous wavelength tuning was obtained between 962.0 nm and 985.5 nm. This tuning has mode-jumps of the order of 115 pm, in accordance with the observed mode-spacing described in Sec. 7.4 and with the cavity length plus a small penetration depth.

Over this tuning range and for a constant amplifier current, the MOPA output power varies between $P_{MOPA} = [480, 1040]$ mW. This is shown in Fig. 8.26(b) where the output power of the individual lasing modes (wavelengths) is plotted. Once again, a mode spacing of 115 pm was observed, with some modes being skipped (e.g. at 965 nm). The observed variation in the output power during the wavelength tuning is likely due to the low I_{RW} current, providing a non-proper seeding of the PA. In addition, the detuning between the emission wavelength of the MO and peak gain wavelength of PA, which for the wide tuning range of the MO also influences the output power during the tuning.

8.5.5 Stability

The stability of this MOPA system was investigated by measuring the output power and emission wavelength for a period of 1 hour, see Fig. 8.27. Over this period, the average output power was (1.070 ± 0.001) W, with a power variation of about 0.11%. Over the same time period, the measured wavelength was 978.79 nm ± 17 pm, with a variation corresponding to the resolution limit of the spectrometer.

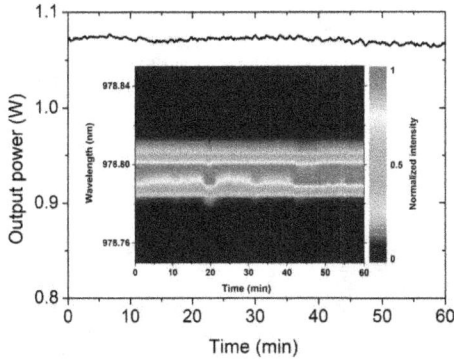

Figure 8.27: *Stability measurement of the output power for a period of 1 hour at* $T = 15°C$. *Inset: corresponding emission wavelength.*

8.5.6 Summary

A compact hybrid MOPA system was presented consisting of a SG-DBR laser and a TPA. The emitted light of the SG-laser was collimated and coupled into the PA using cylindrical micro-lenses. The MOPA system emits an output power of 1040 mW at a wavelength of 979.08 nm, with a spectral width smaller than 20 pm (resolution limited). The emitted light was nearly diffraction limited with a propagation factor of $M_{1/e^2}^2 = 1.6$ along the slow axis.

By operating one of the two heaters placed on top of each SG, 21.1 nm of discrete wavelength tuning could be obtained, with a SG peak-peak spacing of about 2.3 nm. By operating both heaters, a tuning map with 23.5 nm of quasi-continuous tuning was demonstrated with a mode-spacing of about $\delta\lambda = 115$ pm. The tuning range spans between 962.0 nm and 985.5 nm, over which the output power varies between 482 and 1042 mW. This strong power variation is believed to be caused by the low amplifier current I_{RW}, and due to the detuning between the emission wavelengths of the MO and the peak gain wavelength of the PA.

The stability of the system was tested for a period of 1 hour, showing an average output power of 1070 mW ± 1 mW, at a wavelength of 978.79 nm ± 17 pm. Thus no indications of instabilities or feedback effects were observed.

The combination of diffraction limited high output power, tuning, narrow spectral width and compactness makes this device ideal for non-linear frequency conversion applications, in particular upconversion based hyperspectral imaging in the long IR wavelengths.

Future work should consider increasing the output power by reducing instability and self-lasing effects of the PA at higher injection currents. As mentioned, these effects are likely due to the high reflectivity ($\sim 60\%$) of the front SG, whereas the previous MOPA systems utilized MOs with 30% front reflectivity. Self-lasing and instability effects could perhaps be reduced by reducing the front SG reflectivity, or by implementing an optical isolator between the MO and PA. An isolation will reduce the feedback between the PA (front facet) and the MO considerably, and the PA can potentially be operated at higher injection currents with no/limited self-lasing effects. In addition, an automated tuning scheme should be developed which can operate the individual SG heaters and the phase section simultaneously, in order to obtain a fully continuous wavelength tuning.

8.6   Conclusion and discussion

Three different compact MOPA systems were presented in this chapter, all utilizing a TPA in a hybrid configuration. The main results of the three systems are summarizied in Table 8.2.

Table 8.2: *Tuning characteristics of the three MOPA systems.*

MO	$\Delta\lambda$	$\delta\lambda$	P_{max}	ΔP	M^2_{1/e^2}	P_{CL}
type	[nm]	[pm]	[W]	[%]		[%]
DBR-RW	5.5	64	6.2	< 0.2	1.3	79
Y-branch	9.7	52	5.5	< 0.5	2.2	72
SG-DBR	23.5	115	1.0	< 50	1.6	74

The DBR-RW based MOPA system provided 5.5 nm of wavelength tuning, with mode jumps of the order of 64 pm at an output power of 6.2 W. Over this tuning range, an almost constant output power was maintained with a power variation < 0.2%. This small power variation was obtained by seeding the PA above its saturation level to such a level, that even with the power drop of the MO during the wavelength tuning, the PA was still seeded properly.

The Y-branch based MOPA system utilized a Sinusoidal S-bend structure (see chapter 4) as MO, which provided an extended tuning range of 9.7 nm at an output power of 5.5 nm. Once again, by over-seeding the PA, a small power variation < 0.5% was obtained during the wavelength tuning. This system showed the capability of extending the tuning range by implementing multiple-arm lasers, such as the four- or six-arms lasers of chapter 5. However, to fully utilize the extended tuning range, each DBR grating must be optimized according to the tuning range. The larger M^2 value of this system in comparison to the two others is likely explained by a worse coupling into the PA during the glueing process of the micro-lenses.

The MOPA system with a SG-DBR laser provided up to 1.0 W of output power. This low output power (compared to the first two MOPA systems) was explained by the low PA injection current, in particular through the RW section. By increasing the RW current, the MOPA system became unstable and self-lasing effects occured, due to an arising cavity between the front SG facet of the MO and the front facet of the PA. By operating the back SG heater, a discrete tuning of about 21.1 nm was obtained, with a SG peak-peak spacing of about 2.3 nm. By tuning both SGs, a quasi-continuous tuning of 23.5 nm was achieved with a mode spacing of about $\delta\lambda = 115$ nm. Over this tuning range, the output power varied between 0.5 W and 1.0 W, which was explained by a detuning between the MO and PA.

Future work should therefore focus on increasing the output power, by reducing these instability and self-lasing effects. This could e.g. be done by reducing the front SG reflectivity, or by implementing an isolator which will reduce the optical feedback between the PA and MO.

The SG based MOPA system provides the widest tuning, however requires a sophisticated tuning scheme. The alternative concept of multiple-arm seed lasers offers a simpler tuning scheme, however requires a switching mechanism between the laser arms.

Limit and possibilities

For the targeted applications, a hybrid MOPA configuration is versatile and can provide a good understanding of the individual components. In addition, this configuration reduces the thermal feedback effects between the MO and PA. As the PA is injected with a high current, spacing the MO and PA apart reduces the overall thermal feedback. In addition, the micro-heaters of the MO provides a relatively high temperature gradient, which in a monolithic MOPA configuration would cause a reduction in the overall output power (as the PA also gets warmer) during the wavelength tuning.

The three MOPA systems of this work all provide wavelength tuning with mode hops. While this is typically unwanted in many applications, this effect is not an issue in the targeted upconversion applications.

The maximum tuning range depends on the heating capability of the micro-heaters, and is ultimately limited by the width of the gain curve, which depends on the active medium embedded within the vertical structure. While similar MQWs provide higher output powers (as used in the TPAs), the width of the gain curve gets narrower, and thus a trade-off must be made between the maximum output power and tuning range.

Alternatively, by implementing MQWs with varying thickness and/or composition in the active region, the gain profile can be extended as already discussed in chapter 6.

The power variations of the first two MOPA systems during the wavelength tuning, although were very small, are likely explained by the gain shape of the PA and the detuning between the MO and PA. While the latter can be minimized during the design of the MO and PA vertical structures, a power variation is still to be expected due to the gain profile. In fact, one can expect this variation to be larger when extending the tuning range and thus a strong power variation be expected from a wide tuning range.

Finally, a 30% front facet reflectivity was used in the first two MOPA systems as it minimized the optical feedback between the PA and the MO, and provided stable spectral performance. In the case of the SG laser however, the front SG had a maximum 60% reflectivity, which caused the instabilities and self-lasing effects. These effects could be reduced by designing a front SG with 30% reflectivity. Alternatively, applying a micro-isolator between the MO and PA, would reduce the feedback between the PA and the MO.

Chapter 9

Applications

Preliminary results of two upconversion experiments that both utilize the developed MOPA systems are shown in this chapter. The implementation of a MOPA based pump sources into an upconversion experiment is currently work under progress. The presented results serve as a proof of concept of the capabilities that tunable high power MOPA systems offer.

9.1 Upconversion detection

In the following experiment, the DBR-RW based MOPA system was used as a single frequency light source, to efficiently upconvert long IR wavelengths between 9.7 to 10.2 μm. Thus, the motivation behind this experiment is simply to test the laser in long wavelength upconversion experiments.

The setup for the upconversion system is shown in Fig. 9.1. A quantum cascade laser (QCL) from *Block engineering* [107] is used as an IR light source. This laser can be tuned between 9.4 to 12 μm, delivering 50 ns pulses with a time separation of 1 μs. The emitted light from the QCL laser is collimated using two ZnSe cylindrical lenses of focal length $f_1 = 25.4$ mm and $f_2 = 50.8$ mm. A half-wave plate ($\lambda/2$) is then used to align the the polarization of the IR radiation along the extraordinary axis of the non-linear crystal. The collimated IR light is then focused using a lens with a focal length of $f_3 = 50$ mm, inside a AgGaS$_2$ non-linear crystal with a cut angle of $\theta = 43.3°$. A more detailed description of this setup can be found in [50]

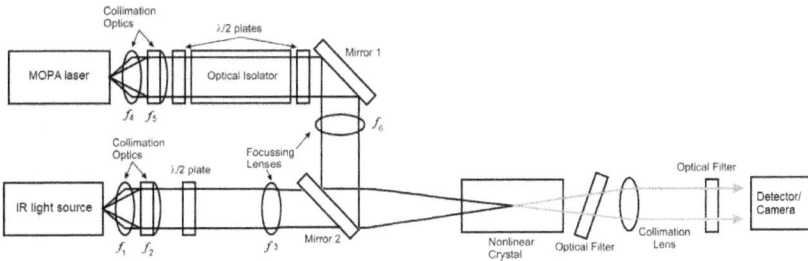

Figure 9.1: *Single-pass upconversion system.*

As mentioned, the pump source is the DBR-RW based MOPA system emitting at 972 nm, which is collimated using an aspheric lens of $f_4 = 3.1$ mm and then a cylindrical lens of $f_5 = 12.7$ mm. The light passes through a half-wave plate before entering an optical isolator. By rotating the half-wave plate before the isolator, the amount of light that enters the system can be controlled. After the isolator, the light passes through another half-wave plate where the polarization of the light is corrected in regards to the non-linear crystal. The pump source is then focused inside the non-linear crystal using a lens with $f_6 = 150$ mm.

The residual pump and IR light are filtered away before the upconverted light is detected using a thermal detector or a CCD camera. Fig. 9.2 shows the upconverted light generated inside the non-linear crystal. Due to the high brightness of the QCL laser, the upconverted light can easily be seen with a CCD camera.

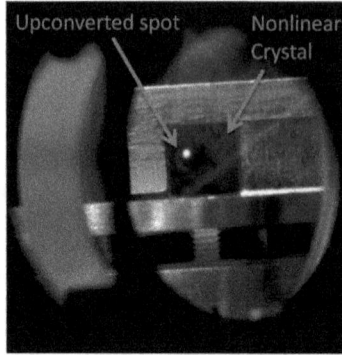

Figure 9.2: *Photograph of the non-linear crystal through the collimation lens, showing a white bright spot corresponding to the upconverted light.*

In this experiment, the QCL wavelength is tuned between 9.4 to 10.2 μm, while keeping the pump wavelength constant at 972 nm. The upconverted light have wavelengths between 880.9 and 887.4 nm and is detected using a thermal detector. The normalized output power of the upconverted light is shown in Fig. 9.3. It is seen that the different IR wavelengths are upconverted with different efficiencies determined by the phase matching condition. Here, a Sinc-like conversion profile is expected, with the peak being centred at the wavelength with the highest conversion efficiency. However, the observed conversion shape corresponds to the QCLs output power variation at the different IR wavelengths as reported in [50], and not due to an absorption effect.

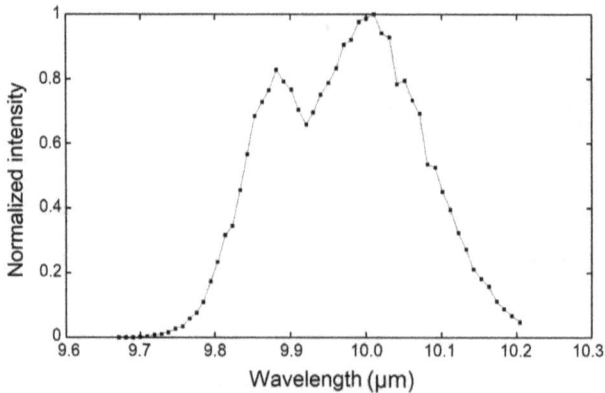

Figure 9.3: *Normalized output power of the upconverted light as function of the IR wavelength.*

The second part of this experiment considers keeping both the pump and the IR wavelenths constants, while rotating the angle of the non-linear crystal between $-2.0°$ to $1.6°$. The normalized output power of the upconverted light is shown in Fig. 9.4. A Sinc-like function is shown as expected from a non-linear frequency conversion process.

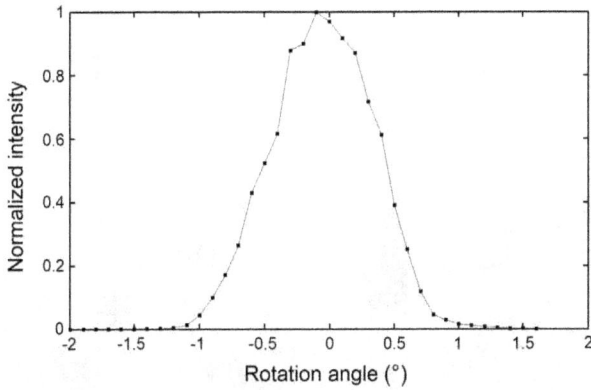

Figure 9.4: *Normalized output power of the upconverted light as function of crystal rotation angle.*

9.2 Hyperspectral Imaging

The second experiment utilizes the tuning capabilities of the MOPA system in an upconversion imaging system. Here, a glowbar is used as a broadband IR light source instead of the QCL. The emitted light of the glowbar undergoes a 4-f imaging system as described in [33]. Once again, a AgGaS$_2$ crystal is used but with a 48° cut angle covering the 6 to 7 μm range.

In this experiment, a CCD camera is used instead of the thermal detector. After aligning the pump laser with the IR light of the glowbar inside the non-linear crystal, a round upconverted spot is seen on the CCD camera, see Fig. 9.5(a). By inserting a Polystyrene film in front of the glowbar, a missing ring is observed, see Fig. 9.5(b). This ring corresponds to an absorption peak of Polystyrene. In addition, by placing both the Polystyrene film and a resolution target in front of the glowbar, both spectral and spatial information can be obtained at the camera, see Fig. 9.5(c).

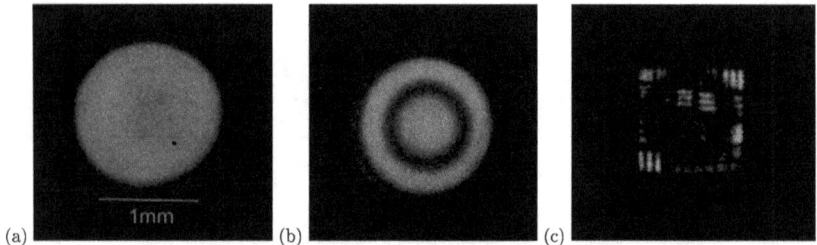

(a)   (b)   (c)

Figure 9.5: *Upconverted images showing (a) the upconverted signal alone (b) the upconverted signal through a polystyrene absorption film and (c) upconverted signal through the absorption film and the resolution target.*

The resolution target is a 1951 USAF [108]. Its structures are seen in Fig. 9.6 next to the upconverted image through it. Using the resolution target, an estimated spatial resolution of 35 μm is obtained from the upconverted image.

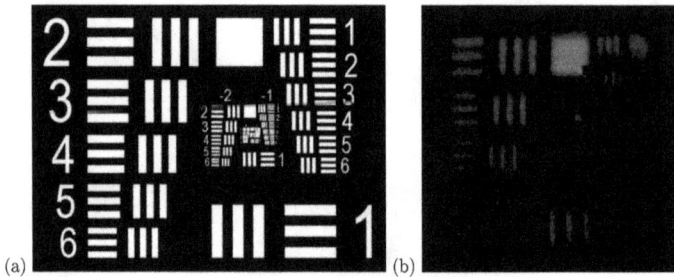

(a)   (b)

Figure 9.6: *(a) 1951 USAF resolution test targets and (b) the corresponding upconverted image through the target.*

By tuning the laser wavelength, different phase matching conditions are fulfilled between the pump wavelength and the broadband IR light of the glowbar. This provide a method for obtaining images of the same sample but at different IR wavelengths; hyperspectral imaging. Fig. 9.5(a)-(c) shows images obtained around 6.3, 6.6 and 6.9 μm, respectively. Polystyrene have strong absorption at these IR wavelengths as shown in Fig. 9.8.

104

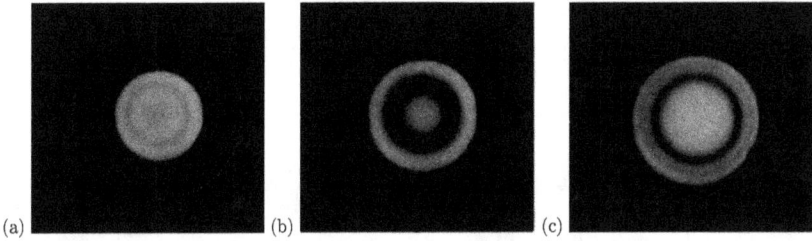

Figure 9.7: *Hyperspectral images obtained at (a) 6.3 (b) 6.6 and (c) at 6.9 µm*

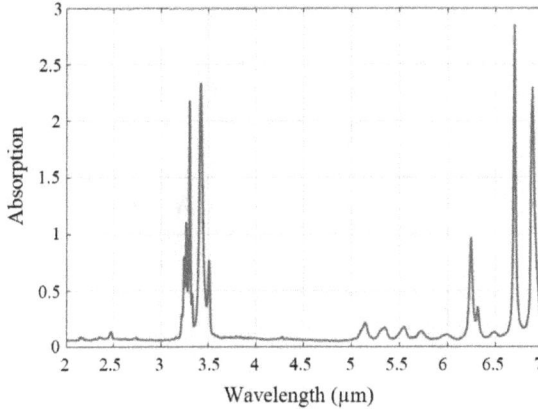

Figure 9.8: *Polystyrene absorption spectra obtained using Fourier-transform infrared spectroscopy (FTIR). This measurement was made at the Technical University of Denmark.*

9.3 Summary

The presented results show the capabilities of the developed MOPA systems. While these are still preliminary results, the combination of high power and wide wavelength tuning seems promising in upconversion applications. One interesting feature that is worth investigating is the pulsing and/or modulation capabilities of the developed MOPA systems. By synchronizing the pump wavelength with a pulsed IR source, upconversion can only occur at a certain time window. This will reduce the upconversion of background thermal noise, and could potentially improve the overall noise performance of the system [109].

Chapter 10

Summary and outlook

The aim of this work is to develop tunable high power laser sources emitting around 976 nm. This was successfully realized through MOPA systems that utilize two different types of tunable lasers. The first approach uses multi-arm DBR-RW lasers, where the introduction of additional arms expands the tuning range in comparison to "standard" DBR-RW lasers. The second approach uses SG structures to obtain wide wavelength tuning from a single laser cavity. These tunable lasers were implemented into MOPA systems that utilizes a TPA structure to provide output powers in the watt range. This was done on miniaturized optical benches that have the dimensions 25 mm × 25 mm. The developed MOPA systems were shown to be suitable for the targeted application, where early-stage upconversion experiments were demonstrated including upconversion detection and hyperspectral imaging.

In the following, the main results of this thesis are summarized and discussed in regards to further development and outlook.

Tunable multi-arm lasers

The first approach of obtaining tunable diode lasers was based on developing tunable multi-arm DBR-RW lasers. These lasers include two-arm (Y-branch), four arm (laser A and B) and a six-arm (laser C), with each arm having its own DBR grating and micro-heater for thermal wavelength tuning. These devices emits around 976 nm with output powers between 200 and 250 mW.

Y-branch DBR-RW lasers

The investigation of the Y-branch lasers includes the study of three different bend structures; a Sin, Cos and a SB based structures. It was shown that the Y-branch structures emit reduced output powers in comparison with straight or S-shaped "single-arm" DBR-RW reference lasers. The Y-branch lasers offer nearly diffraction limited output powers with $M^2_{2.\text{Mom.}}$ values ranging between 2.6 and 1.6. Through numerical simulations and comparison with the reference lasers, this effect was explained by the common section where the two arms are joint. At this point, the RW suddenly become twice as wide and starts to support higher order modes which deteriorates the beam quality.

In addition to the spatial influence of the common section, it was experimentally shown that this section also affects the spectral performance of these lasers. By simply not operating this section, a small absorbent section was obtained which acts as a spectral filter. Spectral multi-mode operation at high output powers was observed when this section was operated, however by setting $I_Y = 0$ mA single mode operation was maintained over the entire investigated current injection range.

The performed simulations of this study were based on calculations of passive RW structures, which could describe the near field behaviour of the Y-branch lasers. However, due to their clear limitations the far field simulations did not provide a complete description. This suggests the development of active cavity simulation tools, which could potentially also describe the spectral dependence of the common section of these lasers.

Among the three studied types of Y-branch lasers, it was seen that the Cos based laser provides the highest output power, in accordance with the simulated transmission coefficients. This favoritizes this device in applications where spectral single-mode high output powers are needed. The Sin based laser offered less output powers, but with improved beam quality that changed the least with increasing output powers. This makes this device suitable for applications with higher beam quality demands. The SB based Y-branch laser offered performance in between the two other lasers, providing similar output powers to the Sin laser, while having comparable beam quality to the Cos laser.

Four-arm DBR-RW lasers

After realizing the strong influence of the common section, four-arm lasers were developed with different common section structures. In laser A the four arms were joint at a single intersection point, while laser B had a joining section for the two inner arms, which was then joint with the two outer arms at a displacement of about 150 μm. In addition, this investigation compared different arm curvatures as the outer arms had stronger bending curvatures than the inner.

The results of this study clearly indicate that the structure in laser B is preferred as it provided comparable laser performance from all four laser arms. This is in contrast to laser A, where reduced output power and spatial performance were observed for the outer arms in comparison to the inner.

Interestingly enough, this mismatch in performance from laser A suggest that a single intersection point acts as a spatial filter for the inner arms, while degrading the performance of the outer arms. In fact, a better beam quality was obtained for the inner arms of laser A in comparison to the four arms of laser B. While this "spatial filtering" effect did degrade the performance of the outer arms, one could use this effect to improve the spatial performance at the intersection points of multi-arm lasers.

Finally, the comparable laser performance of the four arms of laser B indicates, that the different (inner vs. outer) arm curvatures do not affect the laser performance.

Six-arm DBR-RW laser

The influence of the arm curvature was further investigated by developing a six-arm laser with three different arm curvatures, combined through three individual intersection points. This structure provided comparable output power performances with some smaller difference between the six arms. This being said, the outer arms (with the largest curvatures) had worse beam quality when compared to the four inner arms. While these discrepancies are still smaller than those observed in the case of laser A, these results indicate that the bending curvatures must be held below a certain level.

Further development

The results of the multi-arm DBR-RW lasers indicate that relatively high output powers with moderate beam qualities can be obtained. This makes these devices suitable for various applications where lasing at multiple wavelengths is needed. While the non-Gaussian far fields of these devices limits their usage, e.g. in imaging applications, their well defined beam waists and relatively high output powers enables their usage in MOPA systems and in systems with fiber coupling.

The spectral and spatial dependence of the common section between the individual arms suggest, that more emphasis should be made on this section in future developed devices. This includes the usage of multiple intersection points, keeping bending curvatures within a certain limit and having isolated current injection sections. The latter point is important in regards to having spectral single mode operation, where an absorbent common section acts as a spectral filter.

Finally, it is clear from this study that the introduction of additional arms deteriorates the laser performance which is the "cost" of this combination. However, the nearly-diffraction limited beam quality obtained from the Y-branch lasers suggest that multiple Y structures should be combined in multi-arm devices. I.e. having two Y-branch structures connected through a third Y-branch structure, combining in total four arms into a single aperture. This could potentially provide multi-arm laser devices with improved spatial characteristics. Such structures are however likely to become quite long, especially while maintaining the design criteria of $w \ll L$. Having long devices in itself is not an issue during manufacturing process, but it introduces some technical difficulties during the laser mounting and soldering.

Micro-heater wavelength tuning

Thermal wavelength tuning was obtained by embedding resistor based micro-heaters on top of each DBR grating. One of the main findings of this investigation is the extended tuning range, obtained when connecting multiple micro-heaters in series. This configuration provides the highest induced electrical power, due to the increased total resistance and subsequently the widest wavelength tuning. The drawback of this configuration is however the strong output power drop during wavelength tuning, due to the induced large thermal gradient. Therefore, a trade-off must be made between tuning range and output power drop during this tuning.

Such an effect can be reduced by using electrical wavelength tuning, where the injected current causes a slight change in the index of refraction of the grating section. This being said, this is yet to be developed and optimized in GaAs materials. Not only will electrical tuning reduce the output power drop/variation, but potentially also provide higher tuning speeds. Thermal tuning speed is ultimately limited by by the material (GaAs) heat dissipation rate, i.e. how quick the heating/cooling of the material can take place. In the developed devices with the described micro-heaters, a tuning range of about 4.4 nm could be obtained within tens of milliseconds.

The micro-heaters provide quadratic tuning behaviour as function of the injected current due to the Joule heating effect. In addition, quasi-continuous wavelength tuning was observed with a mode spacing set by the cavity length. The investigated devices showed different tuning capabilities due to their different heater and grating dimensions. These results indicate that wider tuning can be obtained from shorter grating structures due to smaller heating surfaces. Heating of smaller surfaces can be made more efficiently if the dimensions of the grating and heaters are adjusted and optimized to one another. This suggest developing shorter grating structures of a certain length which still provide sufficient reflectivity properties.

The main finding of this investigation is that a 7.5 nm of tuning range can be obtained from each arm, demonstrated from a Y-branch DBR-RW laser. This indicates that a potential combined tuning range of $N \times 7.5$ nm can be realized for a laser of N arms, where the individual gratings are spectrally separated by 7.5 nm.

Tuning limits

The multi-arm laser approach provides a combined wider tuning range in comparison to "standard" single arm lasers. This is however limited by the number of arms and ultimately by the available gain bandwidth of the active material. While SQW structures (GaAs based) provide gain bandwidth of tens of nanometres, the usage of MQW structures can provide a wider gain bandwidth. This effect is typically used in superluminescent diodes where MQW structures with varying thickness and/or composition are used. The drawback of "stretching" the gain bandwidth is the reduction in the gain amplitude, which corresponds to a reduced lasing output power. Hence a trade-off must be made between output power and gain bandwidth.

Tunable Sampled-grating lasers

The second approach uses SG structures to obtain wide wavelength tuning from a single laser cavity. In this work, passive and active cavity simulations were presented, with the Transfer matrix model used to design the SGs. The coupled mode equation was then used to simulate the active cavity during wavelength tuning. The presented theory together with these simulation tools provide design approach of SG lasers, having single mode operation over a desired tuning range.

Different vertical structures were fabricated including a SQW structure with a 1st order grating, providing the highest coupling coefficient of $\kappa = 500$ cm^{-1}. The drawback of this structure was however a large far field angle of about $41°$ which limits the usability of these devices. This was compensated by using a DQW based structure, with a 1st order grating and a wider n-cladding layer. This improved structure emits higher output powers, provides a reduced yet suitable coupling coefficient of about 250 cm^{-1}, with a narrower far field angle of $24°$. 2nd order gratings were also developed on the SQW and DQW structures, however they provided a reduced output powers and lower coupling coefficients than the 1st order gratings.

Based on the design parameters and the developed vertical structures, SG laser devices were manufactured providing up to 70 mW of output powers from diffraction limited beams. By operating a single SG heater, up to 21 nm of discrete wavelength could be obtained. When operating both heaters, 23.5 nm of quasi-continuous tuning was achieved. These laser sources serve as the first demonstration of GaAs based SG lasers.

An excellent agreement was observed between the simulated and the measured performance of these devices. This includes the designed reflectivity spectra, tuning behaviour, tuning range and single mode operation.

A SG laser provide wide wavelength tuning from a single cavity, in comparison to the multi-arm laser sources. This being said, a more sophisticated tuning scheme is needed, in particular if a continuous tuning is required. This is in contrast to the simpler tuning scheme of the multi-arm lasers. Another drawback of SG lasers is the two-step MOVPE processes, which increases the risks of contamination between the two steps and requires a cleaning process in between.

Further development

The developed SG lasers were designed to emit single mode over a 23 nm tuning range. This tuning range can be made larger by reducing the period difference between the front and the back SG. E.g. by having a difference of $\Delta\Lambda = 4$ μm, a tuning range of 30 nm can be expected. Obviously, as the tuning range is increased, the system must be adjusted accordingly to maintain single mode operation.

The obvious next step of this work is to apply the phase section to obtain fully continuous wavelength tuning. In particular, this requires a programmable tuning scheme which controls both SG heaters and the phase section to adjust the longitudinal mode position accordingly.

Another interesting approach of further developing SG lasers is to combine SGs with a Y-branch structure, where a SG would be positioned at the front section and with two SGs at each end of the two arms. The front SG would define the center wavelength, while the other two SGs would be spectrally displaced by a $\pm RMS/2$. Such a structure could potentially provide about two times the tuning range obtained from a regular SG laser.

Master oscillator power amplifier systems

Compact hybrid MOPA systems were demonstrated in this work, that combines wavelength tuning from MOs with high output power from TPAs. A hybrid configuration was used instead of a monolithic one for different reasons, particularly as it enables the optimization of the two components separately. In regard to tunable MOPA systems, the motivation was to minimize thermal and optical feedback between the MO and PA. As the micro-heaters induce relatively large temperature gradients, and as the PA is injected with multiple amperes of current, spacing the MO and PA apart reduces the thermal decoupling. Such a feedback is expected from a monolithic device, which could cause strong output power variation during the wavelength tuning and/or drift of the emitted wavelengths.

Separating the MO and PA also reduces the optical feedback as back propagating light, that is not guided through the waveguide structure, is less likely to reach the MO. Another method of reducing optical feedback is to use high (30%) front facet reflectivity of the MOs, which reduces the coupling between the back reflected light from the PA into the MO. This configuration also increases the photon circulation time inside the MO cavity, providing overall improved lasing stability. This being said, a too large reflectivity (60%) caused instabilities as was seen in the case of the SG MOPA system.

The drawback of a hybrid configuration is the need for the coupling between the two components. This was done using micro-lenses that require positioning with high precession, which limits mass production of such devices. Having power variations during the wavelength tuning is perhaps acceptable in some applications, and thus the development of monolithic tunable MOPA systems would still be interesting within the photonic community.

Tapered power amplifiers

A TPA was used to amplify the tunable emission of the MO due to different reasons. TPAs utilize a RW section to define a narrow beam waist, while the tapered section provides high output power. In the case of a MOPA system, the RW structure of these amplifiers provides spatial filtering of the seeded light which is another advantage. These devices can provide tens of watts of output powers from nearly diffraction limited beams.

The MO of this work utilize thermal wavelength tuning, which as already discussed causes a power drop during the tuning process. To avoid a similar power drop in the MOPA systems, the PAs were over-seeded in order to obtain a uniform amplification during the wavelength tuning. The applied TPAs have a saturation power of about 15 mW, and by seeding them with 20 to 30 mW, saturation was maintained even during the ($< 10\%$) power drop during wavelength tuning.

The same discussion of using MQWs can be applied in PA structures, however with more emphasis on the trade-off between the gain bandwidth and amplitude. The TPAs used in this work have a DQW structure with $\sim$ 30 nm of gain bandwidth. This was chosen as it fits the realized tuning ranges of the different MOs. PA structures (and high power diode lasers) utilize MQWs of similar dimension to obtain an increased gain amplitude at the cost of a gain bandwidth shrinkage. Ideally, a PA should have a broad (and flat) gain spectrum providing uniform amplification of all the seeded wavelengths, but in reality a trade-off must be made between those two parameters.

Y-branch based MOPA system

Among the constructed MOPA systems of this work, a high power MOPA system that utilizes a Y-branch DBR-RW laser was described and characterized. It provides emission at two wavelengths: λ_{left} = 973.67 nm and λ_{right} = 975.91 nm with a spectral linewidth below 17 pm. This laser emits 5.5 W of output power, with a M^2_{1/e^2} = 2.2 in the slow axis, resulting in a power in central lobe of about 4 W (72%).

Each of the MO arms provides 7.5 nm of quasi-continuous wavelength tuning with mode hops of the order of 52 pm. As the two DBR gratings were spectrally spaced apart by 2.23 nm, 9.7 nm of combined wavelength tuning could be obtained. Over this tuning range, an almost constant output power was maintained, with power variations of only 0.5% and 0.2% from the left and right arms, respectively. The stability of this device was demonstrated over 1 hour, with power fluctuations below 0.03% and a constant emission wavelength within a resolution of 17 pm.

In addition, the developed MOPA system offers simultaneous dual-wavelength emission and tuning from both arms. During this tuning, single mode operation was maintained, and a power stability $< 0.1\%$ was observed during a 1 hour of period.

SG-based MOPA system

Another MOPA system was demonstrated with a SG laser as MO. By operating a single SG heater, a discrete wavelength tuning of 21.1 nm could be obtained with a SG mode spacing of about 2.3 nm. By operating both SG heaters, 23.5 nm of quasi-continuous wavelength tuning was obtained with a mode spacing of about $\delta\lambda$ = 115 pm. Over this tuning range, a power variation between 0.5 and 1.0 W was observed.

The PA of this system, in particularly the RW section was injected with smaller amount of current (I_{RW} = 5 mA) in comparison to the Y-branch MOPA system (I_{RW} = 300 mA). This was done to avoid self-lasing effects that were seen from the SG-MOPA system at high amplifier currents. This is believed to be caused by a cavity arising between the front facets of the MO and PA. This effect is expected to be stronger in the SG-MOPA system, due to its unique reflectivity spectrum with up to a 60% front reflectivity.

Future work should therefore consider minimizing these self-lasing and instability effects, e.g. by reducing the reflectivity of the front SG or by introducing an optical micro-isolator between the MO and PA. An isolator will considerably reduce the feedback from the PA into the MO, which could potentially lead to higher output powers without wavelength instabilities.

The strong power variation during the wavelength tuning was explained by a detuning between the gain curves of the MO and PA. Therefore, tailored PA structures should be developed, which matches the wide tuning range of the SG lasers, at a certain operation temperature.

Further development

The discussion about applying MQW structures could indeed provide a broader gain bandwidth leading to wider wavelength tuning, with the obvious consequence of reducing the gain amplitude. This approach could still be worth investigating, where the reduced output power could be compensated by using two or multiple cascaded amplification stages. While multiple amplification stages are likely to introduce instabilities, this could be compensated by the use of optical micro isolators between some or all of the different components.

Status of this work

This work targets a "classical" wavelength which is used to pump Erbium-doped fiber amplifiers (EDFA) in optical-telecommunication, and to generate blue laser light by means of SHG. This being said, few high power tunable diode lasers can be found around this wavelength. In this section, the status of this work (tuning range and output power) is compared to other types of tunable diode lasers emitting between 800 and 1100 nm, i.e. GaAs based lasers. This is shown in Fig. 10.1 for lasers with nearly diffraction limited beams, and the corresponding tuning parameters are summarized in Table 10.1.

The Y-branch lasers of this work provide up to 250 mW of output powers with 9.7 nm of combined tuning from both arms. This suggests that a six-arm laser with optimized spectral spacing between its six gratings can provide up to 6×7.5 nm = 45 nm of tuning range. These two cases are shown as black and red stars in Fig. 10.1, respectively. Likewise, the developed SG laser is included in Fig. 10.1 as blue stars, providing 23.5 nm of tuning at an output power of about 70 mW.

In comparison, the widest tuning range of GaAs based light sources is obtained from VCSEL lasers with a MEMS structure. These lasers can provide up to 100 nm of tuning range at up to 850 kHz of sweeping rates. The drawback of these light sources is their low output powers which typically lies ≤ 1 mW.

Another challenger to the presented light sources of this work is the so-called V-cavity laser. These light sources emit narrow linewidths around 850 nm, have a small footprint and offers 11.4 nm of electrically controlled tuning range [110].

The main adversary is the external cavity laser where mechanical movement of a an external grating selects and tunes the wavelength. Fig. 10.1 shows six cases of such configurations (triangles), including three commercially available systems. An external cavity configuration can potentially provide as wide tuning as the total gain bandwidth. The output power depends on the cavity structure, and with a TPA structure a couple of watts can be obtained, see Table 10.1. The drawback of these structures is the mechanical adjustments needed and the typically shoebox size footprint. Although only three commercial lasers are shown in Fig. 10.1, a numerous amount of lasers can be bought at these wavelength, offering somewhat lower output powers and/or narrower tuning ranges.

The three MOPA systems of this work are shown as squares in Fig. 10.1, which provide some of the highest output powers between the shown lasers. This being said, an external grating configuration combined with the TPAs of this work could potentially offer wider tuning with similar output powers.

In conclusion, the presented light sources of this work provide a few hundreds of milliwatts of output power with relatively wide electrically controlled wavelength tuning. With a MOPA configuration, the output power of these light sources can be increased up to the watt level, while maintaining the small footprint and tuning. This makes these light sources suitable for integration into compact optical systems.

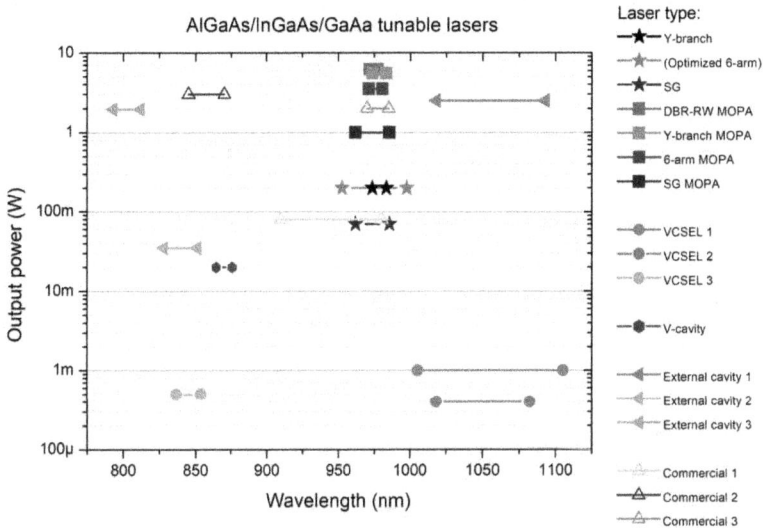

Figure 10.1: *Output power and tuning range of different GaAs based lasers.*

Table 10.1: *Tuning specifications of different GaAa based tunable lasers as indicated in Fig. 10.1.*

Type	Author	Tuning [nm]	Range [nm]	Power
Y-branch	This work	9.7	973.7 - 983.4	250 mW
(Optimized Six-arm)	–	(6 × 7.5)	(950 - 995)	200 mW
SG	–	23.5	962.0 - 985.5	70 mW
DBR-RW MOPA	This Work	5.5	971.8 - 977.3	6.2 W
Y-branch MOPA	–	9.7	973.7 - 983.4	5.5 W
6-arm MOPA	–	9.4	971.3 - 980.7	3.5 W
SG MOPA	–	23.5	962.0 - 985.5	1.0 W
VCSEL 1	Jayaraman et al. [111]	100	1005 - 1105	1 mW
VCSEL 2	John et al. [47]	64	1018 - 1082	0.4 mW
VCSEL 3	Huang et al. [112]	17	837 - 854	0.5 mW
V-cavity	Wei et al. [110]	11.4	858.8 - 881.2	20 mW
External cavity 1	Jensen et al. [45]	75	1018 - 1093	2.5 W
External cavity 2	Chinn et al. [113]	23	828 - 851	35 mW
External cavity 3	Chi et al. [44]	19	793 - 812	1.9 W
Commercial 1	Toptica Photonics [114]	70	910 - 980	80 mW
Commercial 2	Toptica Photonics [115]	25	845 - 870	3 W
Commercial 3	Sacher Lasertechnik [116]	15	970 - 985	2 W

Applications

The demonstration of high power MOPA systems with electrically controlled wavelength tuning is suitable for upconversion applications. This was demonstrated as the wavelength tuning provided a method for obtaining wide IR upconversion from e.g. 6 to 7 μm.

In addition, diode lasers offer pulsing and modulation capabilities which could be interesting to utilize within upconversion applications. Instead of a constant wave (CW) pump source, having a certain time window (synchronized with e.g. a QCL laser) at which upconversion occurs would reduce upconversion of background thermal radiation, leading to an improved noise performance of the system.

The work presented in this thesis is targeted IR upconversion detection, however the developed light sources can also be used in other type of applications. Tunable high power diode lasers are interesting in other non-linear frequency conversion applications such as optical parametric oscillator (OPO). These cavity based laser sources are used to generate two sets of photons; an idler and a signal, both lying in the IR range. Such light sources are typically pumped by fiber lasers emitting around 1 μm, although diode pumped OPOs has been demonstrated [117].

In a similar fashion to upconversion, OPOs are tuned by temperature change or by rotating the non-linear crystal to fulfil different phase matching conditions. In both cases, this is a slow process and in the latter mechanical movement is required. Alternatively, a tunable high power pump source would not only reduce the overall size and complexity of an OPO system, but also offer an easy and quick method for IR wavelength tuning. This being said, the demonstrated output powers of the MOPA systems in this work are still too low, and power scaling is therefore necessary before such an implementation. In addition, it is preferable to have a pump source with continuous tuning to avoid having missing IR regions, which is important in IR absorption spectroscopy applications.

Dual- or even multiple-wavelength tunable laser sources are also interesting in different applications, particularly in tunable THz generation. In such applications, THz radiation can be generated by focusing two laser wavelengths at a photoconductive material that generates THz waves by means of difference frequency generation. Thus compact and tunable THz sources could be realized by using tunable multi-wavelength laser sources.

LIDAR applications would also benefit from having high power lasing emission at multiple near IR wavelengths. For these applications, it would be interesting to investigate the switching speed between the individual MOPA arms, as this will determine how quick a measurement of on/off absorption resonance can be made.

While the tuning range of the demonstrated SG lasers are too narrow and their tuning speeds are too low for high resolution OCT applications, they might still be useful in some low resolution OCT systems. The advantage could be the combination of "moderate" tuning range with high output power from a compact laser source. This being said, OCT systems require continuous wavelength tuning, which could be achieved from a SG lasers by applying the phase section.

Appendix A

Etching depth

The effect of the etching depth of the RW structures used in chapter 3 to 5 will be investigated next. This is done by comparing two (similar) lasers from two different wafers, both based on the vertical structure described in chapter 3. The two wafers differ however by having different RW etching depth; One with a flat (average depth $\approx$ 1232 nm $\pm$ 19 nm) and another with a deep etch ($\approx$ 1372 nm $\pm$ 19 nm), see Fig. A.1. These values were measured using a *P16+* surface profiler from KLA-Tencor with a 0.6 nm resolution [118], and the $\pm$ variation correspond to the SDOM.

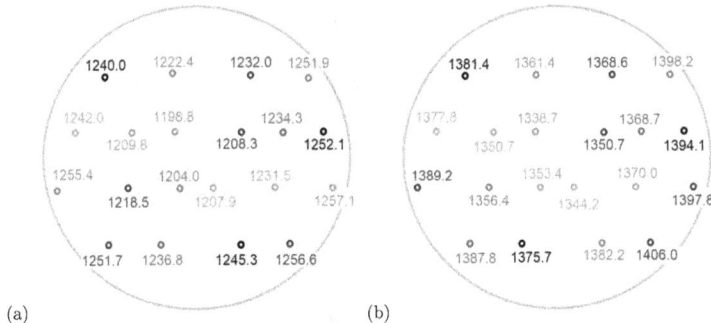

Figure A.1: *Measured RW etching depths in* [nm] *of the flat (a) and deep (b) etched wafers.*

In terms of optical waveguiding, the difference between the flat/deep etching corresponds to a difference in the effective index of refraction Δn_{eff} of the ridge, which is larger for the case of the deeply etched wafer. A larger Δn_{eff} provides a better confinement of the mode within the waveguide structure as described in Sec. 2.2.

This investigation was carried on Y-branch DBR-RW lasers with a Sinusoidal S-bends, see chapter 4. It considers only the left arm of the two lasers, as comparable results were obtained from both the left and the right arms of Y-branch lasers. In addition, in order to keep this investigation simple, only the spatial characteristics are considered in this appendix.

Spatial characteristics

The normalized lateral intensity distributions of the near fields at different injection currents are shown side-by-side in Fig. A.2. Both lasers show "Gaussian-like" near field profiles, with the flat etched laser having a broader side-lope compared to the deeply etched laser.

Likewise, the measured far field profiles are shown side-by-side in Fig. A.3. In the case of the flat etched laser, two "puddles" are seen at low powers which merges together at higher output powers. In the case of the deeply etched laser, a broad high intensity peak was observed with some neighbouring peaks, with a shape that does not change considerably with increasing output powers.

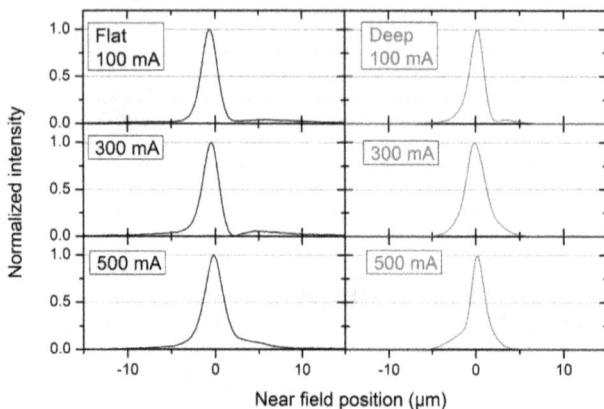

Figure A.2: *Normalized near field profiles of the investigated lasers at different injection currents.*

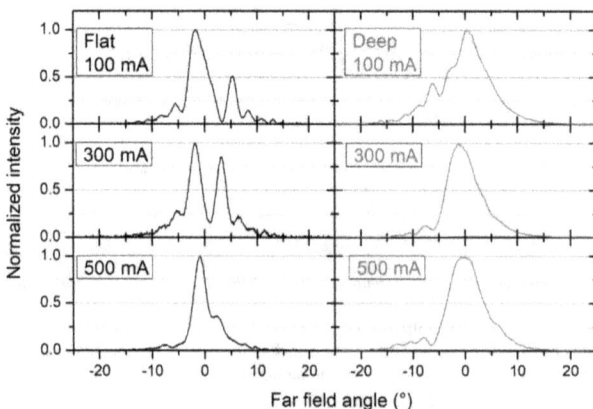

Figure A.3: *Normalized far field profiles of the investigated lasers at different injection currents.*

Summary

Based on this comparison, the deeply etched laser will be the favourite due to its better spatial performance, particularly seen in the far field profiles in Fig. A.3. The overall better beam quality obtained from this laser is explained by the higher Δn_{eff}, which provide a better mode confinement, see Sec. 2.3. Obviously, the better beam quality would be preferential in MOPA systems and in most other applications. Based on this investigation, the described lasers of chapter 3 to 5 were all based on the the deeply etched wafer.

Appendix B

Fitting of the internal parameters

In this appendix, the model used to obtain the internal parameters is described according to [52]. Fig. B.1 to B.4 shows examples of the fittings used to obtain the internal parameters.

The developed vertical structure of chapter 3 was characterized by measuring the threshold current and estimating the slope efficiency under pulsed mode with 1 μs current pulses at a repetition rate of 1 kHz. The differential efficiency η_D was determined using the output power P_0 expression above threshold I_{th}

$$P_0 = \eta_D \frac{h\nu}{q} (I_{\text{inj}} - I_{\text{th}}) \quad for \quad I_{\text{inj}} > I_{\text{th}} \ , \tag{B.1}$$

where h is Planks constant, ν is the frequency, q is the elementary charge and I_{inj} is the injected current. Note that this is under the assumption that the frequency/wavelength is known.

This was done for a 1 mm long laser with a stripe width of $w = 100$ μm; a threshold current $I_{\text{th}} = 158$ mA, a slope efficiency $S = 0.60$ W/A, and a differential efficiency of $\eta_D = 0.94$ were determined. Under the same excitation conditions, the same parameters were determined for devices with $w = 100$ μm width but with 9 different resonator lengths: $L = [400, 600, 800, \cdots, 1800, 2000]$ μm. After obtaining these parameters at different lengths, one can determine the internal quantum efficiency η_i and internal loses α_i using the relationship

$$\frac{1}{\eta_D} = \frac{\alpha_i}{\eta_i \ln\left(\frac{1}{R_f R_b}\right)} L_{cav} + \frac{1}{\eta_i} \ , \tag{B.2}$$

where R_f and R_b are the FM/BM' power reflectivity. For cleaved facet lasers upon GaAs, the product $R_f \times R_b \approx 0.32$. In practice, the reciprocal of the measured differential efficiencies are plotted as function of the laser length L, and fitted with a linear fit. The slope and the intercept of the linear fit can determine/estimate α_i and η_i, respectively. After obtaining the internal losses, one can determine the modal gain Γg_0 according to the relationship

$$\Gamma g_0 = \alpha_i + \frac{1}{L_{cav}} \ln\left(\frac{1}{R_f R_b}\right) = \alpha_i + \alpha_m \ , \tag{B.3}$$

where α_m is the mirror loss. Next, one can construct the modal gain versus the current density as $J_{\text{th}} = \eta_i I_{\text{th}}/(wL)$. The threshold modal gains Γg_{th} can be calculated by

$$\Gamma g_{th} = \alpha_i + \frac{1}{L_{cav}} \ln\left(\frac{1}{R_f R_b}\right) \ , \tag{B.4}$$

for each resonator length L, once the internal loss α_i is found. Thus, it is possible to construct the modal gain versus current density characteristic for the laser from these threshold values. This characteristic is usually fitted well to an exponential of the form

$$J = J_{\text{TR}} \exp\left(\frac{g}{g_0}\right) \ , \tag{B.5}$$

in order to obtain the transparency current density J_{TR} and the gain parameters g_0.

Different parameters depends on temperature, and in general more current is required both for threshold and the increment above threshold as the temperature is increased. The threshold current generally contains three factors which have a significant temperature dependence; the transparency carrier density $N_{\mathrm{tr}} \propto T$, the gain parameter $g_0 \propto 1/T$ and the internal loss $\alpha_i \propto T$. The transparency carrier density is increased and the gain parameter is reduced because the injected carriers spread over a wider range in energy at higher temperatures. The increased internal loss results from the required higher carrier densities for threshold.

Therefore, the threshold current can be approximately modelled by

$$I'_{\mathrm{th}} = I_{th} \exp\left(\frac{\Delta T}{T_0}\right) \, , \tag{B.6}$$

where T_0 is some overall characteristic temperature, and both temperatures are given in degrees Kelvin, K. Note that small values of T_0 indicate a larger dependence on temperature (since $dI_{th}/dT = I_{th}/T_0$).

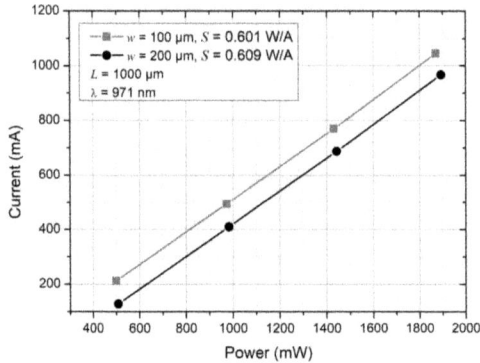

Figure B.1: *Output power as function of the injection current for a 100 and 200 μm wide, 1000 μm long BA laser.*

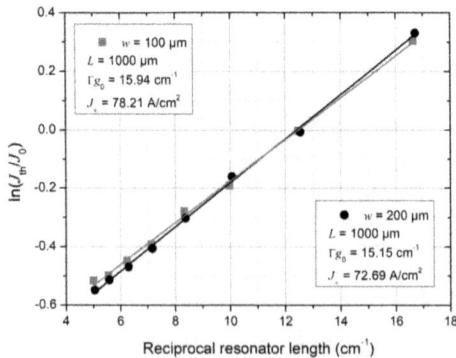

Figure B.2: $\ln\left(J_{th}/J_0\right)$ *as function of reciprocal resonator length for a 100 and 200 μm wide, 1000 μm long BA laser.*

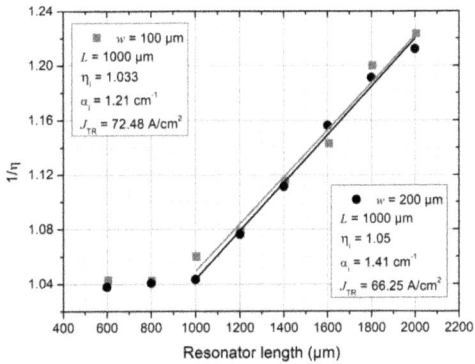

Figure B.3: *Reciprocal differential efficiency as function reciprocal resonator length for a 100 and 200 μm wide, 1000 μm long BA laser.*

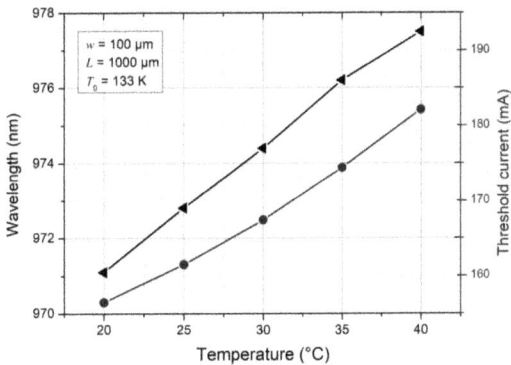

Figure B.4: *Emission wavelength (triangles) and threshold current (circles) at different temperatures, for a 100 μm wide, 1000 μm long BA laser.*

Appendix C

Effective cavity length

The longitudinal mode spacing for a SG-DBR laser can be described as

$$\delta\lambda = \frac{\lambda_0^2}{2n_g L_{cav}} = \frac{\lambda_0^2}{2n_g \left(L_{front,eff} + L_{gain} + L_{phase} + L_{back,eff}\right)},$$ (C.1)

which describes the mode spacing inside a cavity consisting of a gain section and phase section, plus an additional penetration into each SG. In this section, these penetration / effective lengths of the gratings are described.

In the following, consider the simple case of a three mirror laser, consisting of a front facet with reflectivity r_1, a back reflectivity of r_3, and a transition between an active and a passive section with reflectivity r_2, see Fig. C.1.

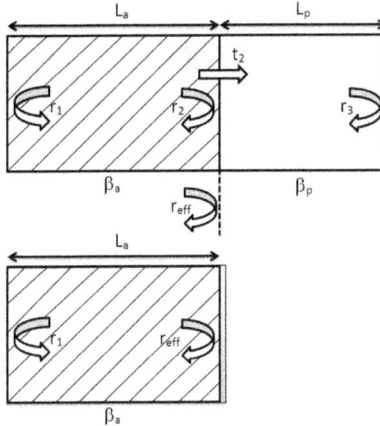

Figure C.1: *Schematic of a three mirror laser (external cavity) with effective mirror to model the external part.*

The transmission coefficient across the interface t_2 includes any scattering or coupling loss, so that in general $r_2^2 + t_2^2 \neq 1$. Fig. C.1 also shows an equivalent two mirror cavity, which replaces the passive section by an effective mirror with reflectivity r_{eff}. The value of r_{eff} can be derived as

$$r_{eff} = r_2 + \frac{r_2^2 r_3 \exp\left(-2i\tilde{\beta}_p L_p\right)}{1 + r_2 r_3 \exp\left(-2i\tilde{\beta}_p L_p\right)}.$$ (C.2)

By implementing eq. (C.2) into the threshold expression one gets

$$\Gamma g_{th} = \alpha_i + \frac{1}{L_a} \ln \left(\frac{1}{r_1 \mid r_{eff} \mid} \right) , \tag{C.3}$$

where Γ and α_i are averages taken over the active section of the cavity only. This is valid as any losses encountered in the passive section are contained in r_{eff}.

Next, the threshold condition is specified for the round-trip phase with

$$r_{eff} = \mid r_{eff} \mid \exp\left(i\phi_{eff}\right) , \tag{C.4}$$

and assumes that both r_{eff} and r_1 are positive and real. The round-trip phase must thus satisfy

$$\exp\left(-2i\beta_a L_a\right) \exp\left(i\phi_{eff}\right) = 1 , \tag{C.5}$$

which translates into $2\beta_a L_a - \phi_{eff} = 2\pi m$. Taking the derivatives of all variables depending on frequency one obtains

$$d\beta_a L_a - \frac{1}{2}d\phi_{eff} = \pi dm . \tag{C.6}$$

The spacing between adjacent modes is found by setting $dm = 1$ and solving for β_a:

$$d\beta_a = \frac{\pi}{L_a - \frac{1}{2}\frac{d\phi_{eff}}{d\beta_a}} . \tag{C.7}$$

In the above expression, the cavity length that defines the mode spacing includes the length of the active section, plus an additional factor depending on how the effective mirror phase changes with frequency. This forms the base for the effective length quantity. As ϕ_{eff} explicitly depends on β_p and not β_a, the effective length is often defined as

$$L_{eff} = -\frac{1}{2}\frac{d\phi_{eff}}{d\beta_p} . \tag{C.8}$$

In the general case of a grating based laser:

$$L_{eff} = -\frac{1}{2}\frac{\partial\phi}{\partial\beta} , \tag{C.9}$$

which can be obtained by different means [52]. In addition, the effective length can be described in terms of the coupling coefficient of the grating [91]

$$L_{eff} = \frac{1}{2\kappa} \tanh\left(\kappa L_g\right) . \tag{C.10}$$

Appendix D

Convergence test

In this chapter, the grid size (resolution) used in the software *Fimmwave* is investigated. Fig. D.1 shows a convergence test of the transmission as function of the applied calculation resolution and the corresponding grid size nx. Convergence is observed at around a resolution of 1 μm, where the variation of the transmission coefficients is of the order of only $6 \times 10^{-3}\%$. This being said, the results shown in chapter 4 were obtained using a resolution of 0.5 μm, just to ensure that the resolution is around half the investigated wavelength. Ideally, the resolution should be at least a quarter of the wavelength, however due to the relatively large structures ($\approx$ 3 mm), this makes the calculations very time consuming.

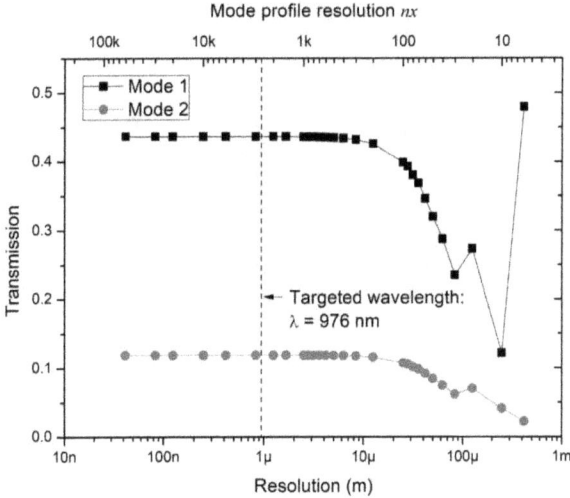

Figure D.1: *Calculated transmission coefficients of the two main modes propagating through a Y-branch structure as function of the applied calculation resolution.*

Appendix E

Photographs of the developed lasers

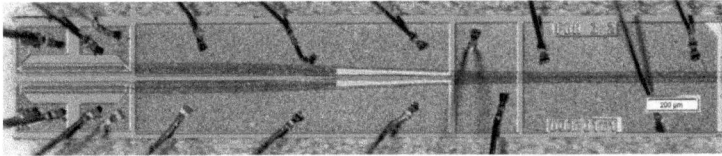

Figure E.1: *Photograph of a Y-branch laser with two micro-heaters (farthest left) above the grating section.*

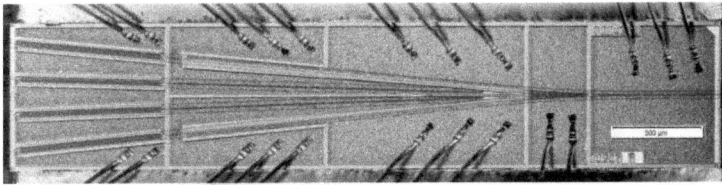

Figure E.2: *Photograph of a four-arm laser.*

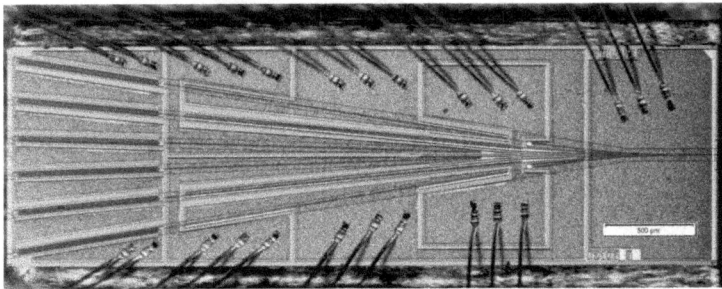

Figure E.3: *Photograph of a six-arm laser.*

Figure E.4: *Photograph of the DBR-RW based MOPA system.*

Figure E.5: *Photograph of the Y-branch based MOPA system.*

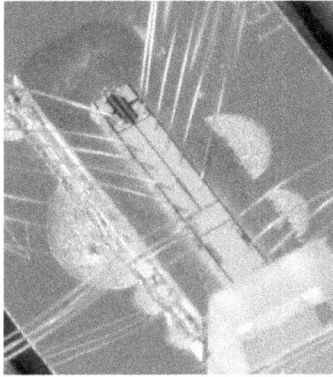

Figure E.6: *Photograph of the bonding scheme for the current injection of the Y-branch laser.*

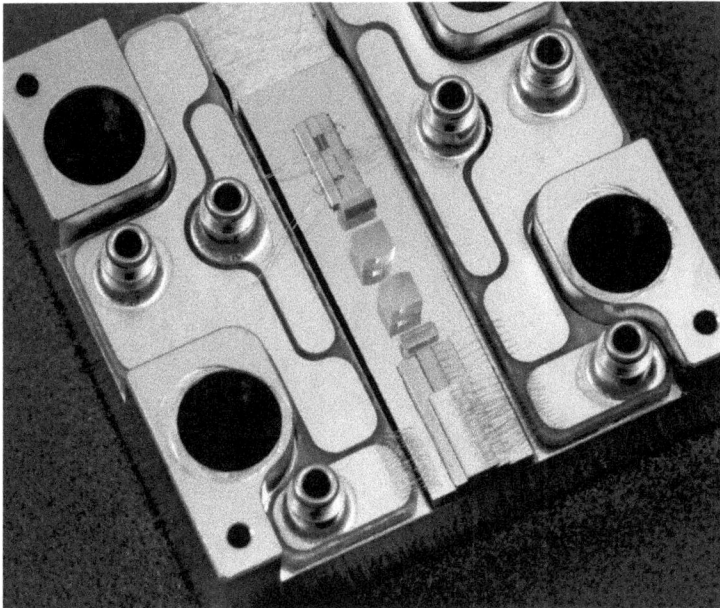

Figure E.7: *Photograph of the SG based MOPA system.*

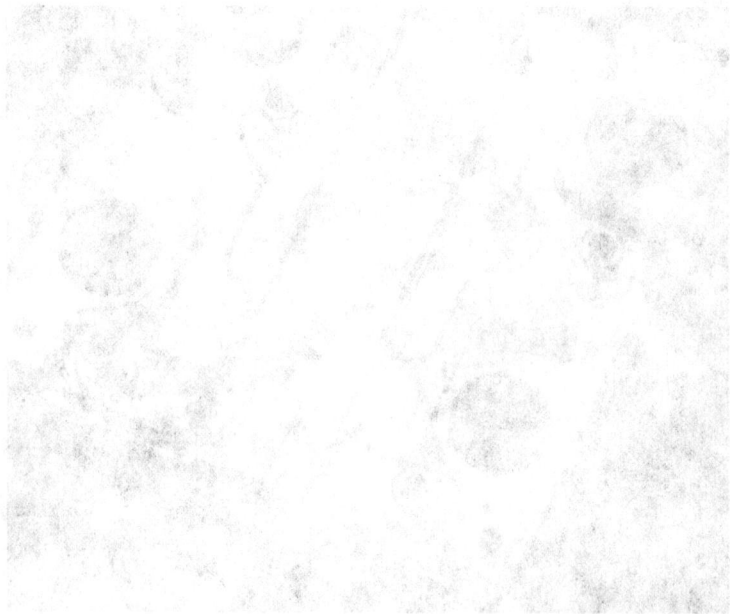

Appendix F

MATLAB code of the Transfer matrix model

In the following, the MATLAB function used to obtain the SG reflectivity spectra is shown. It requires three input parameters: number of periods m, grating length g and period length p. The function returns six parameters: the reflectivity R, the wavelength axis lambda, the maximum reflectivity Rmax, effective length Leff, the propagation parameter b and the phase.

```
function [R,lambda,Rmax,Leff,b,phase] = P_vs_g(m,g,p)
figure(1)
hold on

%% variables
lambda_s=0.950; % wavelength axis start
lambda_e=1.0;   % wavelength axis end
nlambda=10000;  % wavelength resolution
lambda=zeros(1,nlambda);
R=lambda; Q=lambda;
    z_grating=g;
for i=1:nlambda;
lambda(i)=lambda_s+i*(lambda_e-lambda_s)/nlambda;

%% constants
lambda0=0.976;
alpha0=0;
n0=3.424;
ng=4;
n=n0+(n0-ng)/lambda0*(lambda(i)-lambda0);

%% Propagation parameter beta
b(i)=2*pi*n./lambda(i)+1i*alpha0/2;
b_0=2*pi*n0/lambda0;
Delta_b=b(i)-b_0;

%% Coupling coefficient kappa [1/cm]
phi=0;
kappa=200/1e4; % Coupling coefficient kappa [1/cm]
kappa_p=kappa*exp(-1i*2*pi*phi);
kappa_m=kappa*exp(1i*2*pi*phi);

%% T matrix for sections with grating: kappa not 0
y=sqrt((Delta_b).^2-kappa_p.*kappa_m);
T_grating=[cos(y*z_grating)-1i*Delta_b.*sin(y*z_grating)./y,-1i*kappa_p.*sin(y*z_grating)./
```

```
y;1i*kappa_m.*sin(y*z_grating)./y,cos(y*z_grating)+1i*Delta_b.*sin(y*z_grating)./y];

%% T matrix for section without grating: kappa=0
z_no_grating=p-z_grating; % length of grating
phi=0; %b*z_no_grating;
kappa=0; % 1/cm
kappa_p=kappa*exp(-1i*2*pi*phi);
kappa_m=kappa*exp(1i*2*pi*phi);
y=sqrt((Delta_b).^2-kappa_p.*kappa_m);
if lambda(i)==lambda0
    T_no_grating=[1,0;0,1];
else
T_no_grating=[cos(y*z_no_grating)-1i*Delta_b.*sin(y*z_no_grating)./
y,-1i*kappa_p.*sin(y*z_no_grating)./y;1i*kappa_m.*sin(y*z_no_grating)./
y,cos(y*z_no_grating)+1i*Delta_b.*sin(y*z_no_grating)./y];
end

%% Combined T matrix
T=(T_grating*T_no_grating)^m;
phase(i)=angle(T(1,2)/T(2,2));
tt(i)=abs(T(1,2)/T(2,2))^2;
end

%% Reflectivity
R=tt;
Rmax=max(R)*100;
lambda=lambda*1000;

%% Plot the reflectivity spectrum
figure(1)
hold on
plot(lambda,R)

%% Phase
phase=transpose(phase);
b=transpose(b);
R=transpose(R); lambda=transpose(lambda);

%% Effective length
Leff1=-1/2*diff(phase)./diff(b);
x=find(lambda==976);
Leff=Leff1(x);
```

Appendix G

Sampled-grating reflectivity models

In this chapter, a comparison is made between the applied Transfer matrix (TM) model to simulate the reflectivity spectra and the analytical SG model by Jayaraman et. al which is based on a Fourier component analysis. This comparison does not illustrate the analytical model as it is described elsewhere [91]. Instead it simply shows why the TM model is preferred in cases where the coupling coefficients exceeds $\kappa > 200$ cm^{-1}.

By using the two methods, the simulated reflectivity spectra were obtained for a $\Lambda_s = 50$ μm long period, with a grating of length $L_g = 10$ μm, having $m = 20$ periods for $\kappa = 200$ cm^{-1}, see Fig. G.1. As can be seen, the analytical model provides unrealistic reflectivity values above 100%.

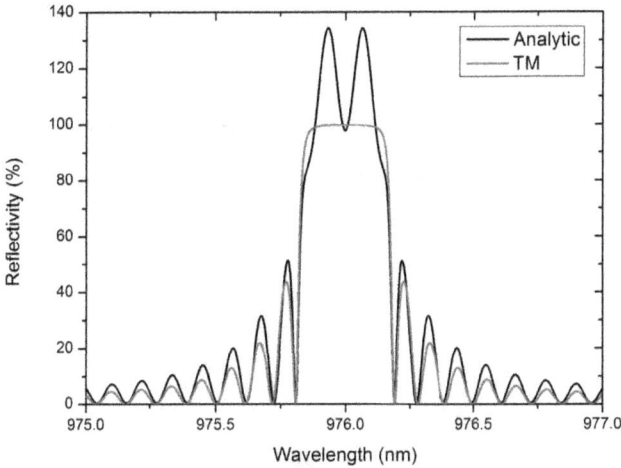

Figure G.1: *Simulated peak reflectivity using the analytic and the TM model.*

The maximum reflectivity for $\kappa = [100/200/300]$ cm^{-1} at different grating lengths is shown in Fig. G.2. As can be seen, higher coupling coefficients provide peak reflectivities > 1, and this deviation is increased even further for longer grating lengths.

Finally, the simulated FWHM using the two models is considered. This is shown in Fig. G.2 for the three different coupling coefficients. The TM model provides linear dependence between the FMWH and the grating length. In the case of the analytic model, a deviation is observed at some grating lengths.

These results suggests that the analytic model provides correct reflectivity values for a certain range of SG parameters. In comparison, the TM model provides realistic reflectivity values even for large coupling coefficients. Therefore, the TM was chosen to simulate the reflectivity spectra shown in chapter 7. The advantage of the analytic model however, is the simple expressions it provides for the different SG design parameters such as; peak reflectivity, the bandwidth of a certain peak, the penetration depth ect.

Period = Grating length + Sampling period = 50 µm, 20 periods

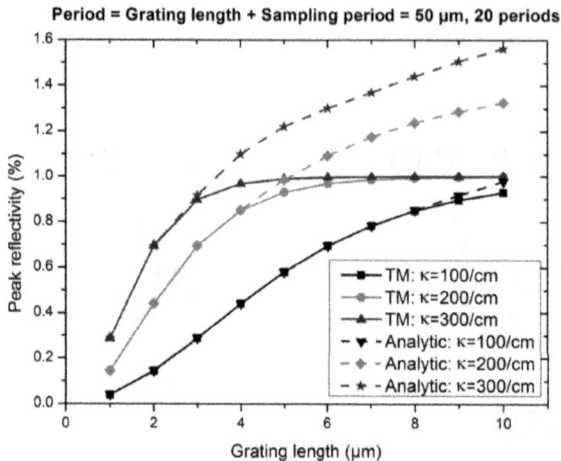

Figure G.2: *Simulated peak reflectivity (at 976 nm) as function of the grating length, for* $\kappa = [100/200/300]$ cm^{-1}, *obtained using the TM and the analytic model.*

Period = Grating length + Sampling period = 50 µm, 20 periods

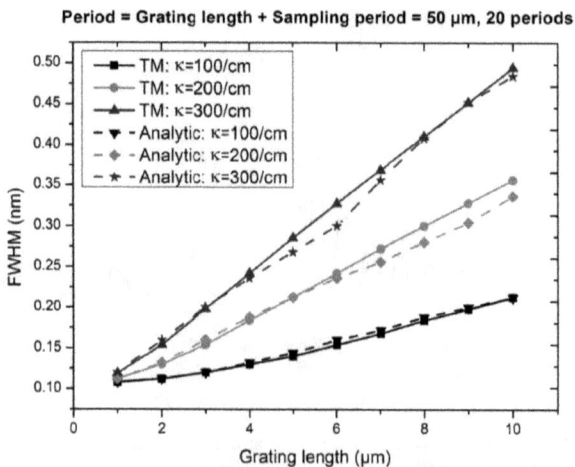

Figure G.3: *Simulated FWHM of the main reflectivity peak (at 976 nm) as function of the grating length, for* $\kappa = [100/200/300]$ cm^{-1}, *obtained using the TM and the analytic model.*

Appendix H

Facet coating

In this section, some of the measured facet reflectivities are shown. Fig. H.1 shows an AR back and a 30% front facet coating used for the DBR-RW and the Y-branch lasers that were implemented into MOPA systems in chapter 8. Likewise, Fig. H.2 shows the AR coating of the front/back facets a SG-DBR laser. These devices were AR coated to minimize Fabry–Pérot modes and to maximize the effects of the developed SGs in chapter 7.

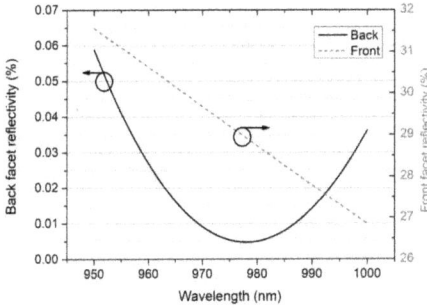

Figure H.1: *Reflectivity curves of the back (left) and front (right) facets at different wavelengths used for the MOs.*

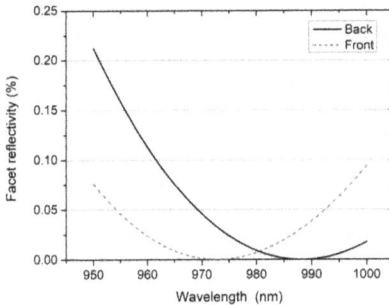

Figure H.2: *Reflectivity curves of the back and front facets at different wavelengths used for the SG-DBR lasers.*

Bibliography

[1] T. H. Maiman, "Stimulated Optical Radiation in Ruby," *Nature*, vol. 187, pp. 493–494, Aug. 1960.

[2] R. N. Hall, G. E. Fenner, J. D. Kingsley, T. J. Soltys, and R. O. Carlson, "Coherent Light Emission From GaAs Junctions," *Physical Review Letters*, vol. 9, pp. 366–368, Nov. 1962.

[3] M. I. Nathan, W. P. Dumke, G. Burns, F. H. Dill, and G. Lasher, "STIMULATED EMISSION OF RADIATION FROM GaAs p-n JUNCTIONS," *Applied Physics Letters*, vol. 1, pp. 62–64, Nov. 1962.

[4] Z. I. Alferov, "The history and future of semiconductor heterostructures," *Semiconductors*, vol. 32, pp. 1–14, Jan. 1998.

[5] R. Diehl, *High-Power Diode Lasers: Fundamentals, Technology, Applications*. Springer Science & Business Media, July 2003.

[6] L. A. Coldren, G. A. Fish, Y. Akulova, J. S. Barton, L. Johansson, and C. W. Coldren, "Tunable semiconductor lasers: A tutorial," *Journal of Lightwave Technology*, vol. 22, pp. 193–202, Jan. 2004.

[7] O. E. DeLange, "Wide-band optical communication systems: Part II Frequency-division multiplexing," *Proceedings of the IEEE*, vol. 58, pp. 1683–1690, Oct. 1970.

[8] B. Glance, U. Koren, C. A. Burrus, and J. D. Evankow, "Discretely-tuned N-frequency laser for packet switching applications based on WDM," *Electronics Letters*, vol. 27, pp. 1381–1383, July 1991.

[9] N. Shibata, O. Ishida, Y. Tada, S. Kuwano, Y. Tohmori, and S. Suzuki, "Performance of four channel FDM crossconnect switching without bit loss," *Electronics Letters*, vol. 29, pp. 800–801, Apr. 1993.

[10] S. Marschall, T. Klein, W. Wieser, B. R. Biedermann, K. Hsu, K. P. Hansen, B. Sumpf, K.-H. Hasler, G. Erbert, O. B. Jensen, C. Pedersen, R. Huber, and P. E. Andersen, "Fourier domain mode-locked swept source at 1050 nm based on a tapered amplifier," *Optics Express*, vol. 18, p. 15820, July 2010.

[11] I. Grulkowski, J. J. Liu, B. Potsaid, V. Jayaraman, C. D. Lu, J. Jiang, A. E. Cable, J. S. Duker, and J. G. Fujimoto, "Retinal, anterior segment and full eye imaging using ultrahigh speed swept source OCT with vertical-cavity surface emitting lasers," *Biomedical Optics Express*, vol. 3, pp. 2733–2751, Nov. 2012.

[12] H. W. Siesler, Y. Ozaki, S. Kawata, and H. M. Heise, *Near-Infrared Spectroscopy: Principles, Instruments, Applications*. John Wiley & Sons, July 2008.

[13] J. T. C. Liu, J. B. Jeffries, and R. K. Hanson, "Wavelength modulation absorption spectroscopy with 2f detection using multiplexed diode lasers for rapid temperature measurements in gaseous flows," *Applied Physics B*, vol. 78, pp. 503–511, Feb. 2004.

[14] K. H. Levin and C. L. Tang, "Wavelength-modulation Raman spectroscopy," *Applied Physics Letters*, vol. 33, pp. 817–819, Nov. 1978.

[15] A. C. De Luca, M. Mazilu, A. Riches, C. S. Herrington, and K. Dholakia, "Online Fluorescence Suppression in Modulated Raman Spectroscopy," *Analytical Chemistry*, vol. 82, pp. 738–745, Jan. 2010.

[16] B. J. Finlayson-Pitts and J. N. P. Jr, *Chemistry of the Upper and Lower Atmosphere: Theory, Experiments, and Applications.* Academic Press, Nov. 1999.

[17] M. Chi, O. B. Jensen, and P. M. Petersen, "Dual-wavelength high-power diode laser system based on an external-cavity tapered amplifier with tunable frequency difference," *Journal of the Optical Society of America B*, vol. 29, pp. 2617–2621, Sept. 2012.

[18] M. Tawfieq, O. B. Jensen, A. K. Hansen, B. Sumpf, K. Paschke, and P. E. Andersen, "Efficient generation of 509nm light by sum-frequency mixing between two tapered diode lasers," *Optics Communications*, vol. 339, pp. 137–140, Mar. 2015.

[19] M. Chi, O. B. Jensen, and P. M. Petersen, "High-power dual-wavelength external-cavity diode laser based on tapered amplifier with tunable terahertz frequency difference," *Optics Letters*, vol. 36, pp. 2626–2628, July 2011.

[20] J. O. Gwaro, C. Brenner, B. Sumpf, A. Klehr, J. Fricke, and M. R. Hofmann, "Terahertz frequency generation with monolithically integrated dual wavelength distributed Bragg reflector semiconductor laser diode," *IET Optoelectronics*, vol. 11, no. 2, pp. 49–52, 2017.

[21] T. N. Vu, A. Klehr, B. Sumpf, H. Wenzel, G. Erbert, and G. Tränkle, "Wavelength stabilized ns-MOPA diode laser system with 16 W peak power and a spectral line width below 10 pm," *Semiconductor Science and Technology*, vol. 29, no. 3, p. 035012, 2014.

[22] T. N. Vu, A. Klehr, B. Sumpf, H. Wenzel, G. Erbert, and G. Tränkle, "Tunable 975 nm nanosecond diode-laser-based master-oscillator power-amplifier system with 16.3 W peak power and narrow spectral linewidth below 10 pm," *Optics Letters*, vol. 39, pp. 5138–5141, Sept. 2014.

[23] J. S. Dam, P. Tidemand-Lichtenberg, and C. Pedersen, "Room-temperature mid-infrared single-photon spectral imaging," *Nature Photonics*, vol. 6, pp. 788–793, Nov. 2012.

[24] M. Hermes, R. B. Morrish, L. Huot, L. Meng, S. Junaid, J. Tomko, G. R. Lloyd, W. T. Masselink, P Tidemand-Lichtenberg, C. Pedersen, F. Palombo, and N. Stone, "Mid-IR hyperspectral imaging for label-free histopathology and cytology," *Journal of Optics*, vol. 20, no. 2, p. 023002, 2018.

[25] D. Townsend, M. Miljković, B. Bird, K. Lenau, O. Old, M. Almond, C. Kendall, G. Lloyd, N. Shepherd, H. Barr, N. Stone, and M. Diem, "Infrared micro-spectroscopy for cyto-pathological classification of esophageal cells," *Analyst*, vol. 140, pp. 2215–2223, Mar. 2015.

[26] A. B. Seddon, T. M. Benson, S. Sujecki, N. Abdel-Moneim, Z. Tang, D. Furniss, L. Sojka, N. Stone, N. Jayakrupakar, G. R. Lloyd, I. Lindsay, J. Ward, M. Farries, P. M. Moselund, B. Napier, S. Lamrini, U. Møller, I. Kubat, C. R. Petersen, and O. Bang, "Towards the mid-infrared optical biopsy," in *SPIE Proceedings*, Optical Biopsy XIV: Toward Real-Time Spectroscopic Imaging and Diagnosis, vol. 9703, p. 970302, Mar. 2016.

[27] A. Rogalski, "HgCdTe infrared detector material: History, status and outlook," *Reports on Progress in Physics*, vol. 68, no. 10, p. 2267, 2005.

[28] M. B. Reine, "HgCdTe photodiodes for IR detection: A review," in *Photodetectors: Materials and Devices VI*, vol. 4288, pp. 266–278, International Society for Optics and Photonics, June 2001.

[29] M. Mancinelli, A. Trenti, S. Piccione, G. Fontana, J. S. Dam, P. Tidemand-Lichtenberg, C. Pedersen, and L. Pavesi, "Mid-infrared coincidence measurements on twin photons at room temperature," *Nature Communications*, vol. 8, p. 15184, May 2017.

[30] L. Høgstedt, J. S. Dam, A.-L. Sahlberg, Z. Li, M. Aldén, C. Pedersen, and P. Tidemand-Lichtenberg, "Low-noise mid-IR upconversion detector for improved IR-degenerate four-wave mixing gas sensing," *Optics Letters*, vol. 39, p. 5321, Sept. 2014.

[31] A. Barh, P. Tidemand-Lichtenberg, and C. Pedersen, "Thermal noise in mid-infrared broadband upconversion detectors," *Optics Express*, vol. 26, pp. 3249–3259, Feb. 2018.

[32] L. M. Kehlet, P. Tidemand-Lichtenberg, J. S. Dam, and C. Pedersen, "Infrared upconversion hyperspectral imaging," *Optics Letters*, vol. 40, p. 938, Mar. 2015.

[33] S. Junaid, J. Tomko, M. P. Semtsiv, J. Kischkat, W. T. Masselink, C. Pedersen, and P. Tidemand-Lichtenberg, "Mid-infrared upconversion based hyperspectral imaging," *Optics Express*, vol. 26, pp. 2203–2211, Feb. 2018.

[34] L. Høgstedt, A. Fix, M. Wirth, C. Pedersen, and P. Tidemand-Lichtenberg, "Upconversion-based lidar measurements of atmospheric CO2," *Optics Express*, vol. 24, pp. 5152–5161, Mar. 2016.

[35] L. Meng, A. Fix, M. Wirth, L. Høgstedt, P. Tidemand-Lichtenberg, C. Pedersen, and P. J. Rodrigo, "Upconversion detector for range-resolved DIAL measurement of atmospheric CH4," *Optics Express*, vol. 26, pp. 3850–3860, Feb. 2018.

[36] R. W. Boyd, *Nonlinear Optics*. Amsterdam ; Boston: Academic Press, 3rd ed ed., 2008.

[37] Q. Hu, J. Seidelin Dam, C. Pedersen, and P. Tidemand-Lichtenberg, "High-resolution mid-IR spectrometer based on frequency upconversion," *Optics Letters*, vol. 37, p. 5232, Dec. 2012.

[38] R. Tang, W. Wu, X. Li, H. Pan, H. Zeng, and E. Wu, "Low-Noise Infrared Spectroscopy via Tunable Frequency Upconversion at Single-Photon Level," *IEEE Photonics Technology Letters*, vol. 27, pp. 1642–1645, Aug. 2015.

[39] R. T. Thew, H. Zbinden, and N. Gisin, "Tunable upconversion photon detector," *Applied Physics Letters*, vol. 93, p. 071104, Aug. 2008.

[40] L. A. Coldren, "Monolithic tunable diode lasers," *IEEE Journal of Selected Topics in Quantum Electronics*, vol. 6, pp. 988–999, Nov. 2000.

[41] J. Buus, M.-C. Amann, and D. J. Blumenthal, *Tunable Laser Diodes and Related Optical Sources*. IEEE, 2 ed., 2005.

[42] M. Wilkens, H. Wenzel, J. Fricke, A. Maaßdorf, P. Ressel, S. Strohmaier, A. Knigge, G. Erbert, and G. Tränkle, "High-Efficiency Broad-Ridge Waveguide Lasers," *IEEE Photonics Technology Letters*, vol. 30, pp. 545–548, Mar. 2018.

[43] K. Paschke, B. Sumpf, F. Dittmar, G. Erbert, R. Staske, H. Wenzel, and G. Tränkle, "Nearly diffraction limited 980-nm tapered diode lasers with an output power of 7.7 W," *IEEE Journal of Selected Topics in Quantum Electronics*, vol. 11, pp. 1223–1227, Sept. 2005.

[44] M. Chi, O. B. Jensen, J. Holm, C. Pedersen, P. E. Andersen, G. Erbert, B. Sumpf, and P. M. Petersen, "Tunable high-power narrow-linewidth semiconductor laser based on an external-cavity tapered amplifier," *Optics Express*, vol. 13, no. 26, p. 10589, 2005.

[45] O. B. Jensen, B. Sumpf, G. Erbert, and P. M. Petersen, "Widely Tunable High-Power Tapered Diode Laser at 1060 nm," *IEEE Photonics Technology Letters*, vol. 23, pp. 1624–1626, Nov. 2011.

[46] T. Ansbaek, I. S. Chung, E. S. Semenova, and K. Yvind, "1060-nm Tunable Monolithic High Index Contrast Subwavelength Grating VCSEL," *IEEE Photonics Technology Letters*, vol. 25, pp. 365–367, Feb. 2013.

[47] D. D. John, C. B. Burgner, B. Potsaid, M. E. Robertson, B. K. Lee, W. J. Choi, A. E. Cable, J. G. Fujimoto, and V. Jayaraman, "Wideband Electrically Pumped 1050-nm MEMS-Tunable VCSEL for Ophthalmic Imaging," *Journal of Lightwave Technology*, vol. 33, pp. 3461–3468, Aug. 2015.

[48] "Introduction to Light Emitting Diode Technology and Applications." https://www.crcpress.com/Introduction-to-Light-Emitting-Diode-Technology-and-Applications/Held/p/book/9781420076622, Dec. 2008.

[49] W. Gandrud, G. Boyd, and E. Buehler, "Phase-matched upconversion of 10.6-μ radiation in ZnGeP2," *IEEE Journal of Quantum Electronics*, vol. 7, pp. 307–308, June 1971.

[50] Y.-P. Tseng, C. Pedersen, and P. Tidemand-Lichtenberg, "Upconversion detection of long-wave infrared radiation from a quantum cascade laser," *Optical Materials Express*, vol. 8, pp. 1313–1321, May 2018.

[51] G. Herzberg, *Molecular Spectra and Molecular Structure*. Van Nostrand, 1950.

[52] L. A. Coldren, S. W. Corzine, and M. L. Mashanovitch, *Diode Lasers and Photonic Integrated Circuits*. John Wiley & Sons, Mar. 2012.

[53] W. T. Tsang, "Extremely low threshold (AlGa)As graded-index waveguide separate-confinement heterostructure lasers grown by molecular beam epitaxy," *Applied Physics Letters*, vol. 40, pp. 217–219, Feb. 1982.

[54] Y. Kotaki and H. Ishikawa, "Spectral characteristics of a three-section wavelength-tunable DBR laser," *IEEE Journal of Quantum Electronics*, vol. 25, pp. 1340–1345, June 1989.

[55] S. D. Hersee, B. de Cremoux, and J. P. Duchemin, "Some characteristics of the GaAs/GaAlAs graded-index separate-confinement heterostructure quantum well laser structure," *Applied Physics Letters*, vol. 44, pp. 476–478, Mar. 1984.

[56] P. Crump, G. Blume, K. Paschke, R. Staske, A. Pietrzak, U. Zeimer, S. Einfeldt, A. Ginolas, F. Bugge, K. Häusler, P. Ressel, H. Wenzel, and G. Erbert, "20W continuous wave reliable operation of 980nm broad-area single emitter diode lasers with an aperture of 96um," in *High-Power Diode Laser Technology and Applications VII*, vol. 7198, p. 719814, International Society for Optics and Photonics, Feb. 2009.

[57] G. P. Agrawal and N. K. Dutta, *Semiconductor Lasers*. Springer US, 2 ed., 1993.

[58] H. Wenzel, F. Bugge, M. Dallmer, F. Dittmar, J. Fricke, K. H. Hasler, and G. Erbert, "Fundamental-Lateral Mode Stabilized High-Power Ridge-Waveguide Lasers With a Low Beam Divergence," *IEEE Photonics Technology Letters*, vol. 20, pp. 214–216, Feb. 2008.

[59] A. Knigge, G. Erbert, J. Jonsson, W. Pittroff, R. Staske, B. Sumpf, M. Weyers, and G. Tränkle, "Passively cooled 940 nm laser bars with 73% wall-plug efficiency at 70 W and 25/spl deg/C," *Electronics Letters*, vol. 41, pp. 250–251, Mar. 2005.

[60] M. Kanskar, T. Earles, T. J. Goodnough, E. Stiers, D. Botez, and L. J. Mawst, "73% CW power conversion efficiency at 50 W from 970 nm diode laser bars," *Electronics Letters*, vol. 41, pp. 245–247, Mar. 2005.

[61] B. Sumpf, M. Maiwald, A. Müller, A. Ginolas, K. Hausler, G. Erbert, and G. Tränkle, "Reliable Operation for 14500 h of a Wavelength-Stabilized Diode Laser System on a Microoptical Bench at 671 nm," *IEEE Transactions on Components, Packaging and Manufacturing Technology*, vol. 2, pp. 116–121, Jan. 2012.

[62] H. Kogelnik and C. V. Shank, "Stimulated emission in a periodic structure," *Applied Physics Letters*, vol. 18, pp. 152–154, Feb. 1971.

[63] B. E. A. Saleh and M. C. Teich, *Fundamentals of Photonics*. Wiley, Mar. 2007.

[64] O. Brox, H. Wenzel, P. Della Casa, M. Tawfieq, B. Sumpf, M. Weyers, and A. Knigge, "Reflectors and tuning elements for widely-tunable GaAs-based sampled grating DBR lasers," in *Novel In-Plane Semiconductor Lasers XVII*, vol. 10553, p. 1055310, International Society for Optics and Photonics, Feb. 2018.

[65] H. Kressel, H. F. Lockwood, and F. Z. Hawrylo, "Large-Optical-Cavity (AlGa) As-GaAs Heterojunction Laser Diode: Threshold and Efficiency," *Journal of Applied Physics*, vol. 43, pp. 561–567, Feb. 1972.

[66] D. Botez, "Cw high-power single-mode operation of constricted double-heterojunction AlGaAs lasers with a large optical cavity," *Applied Physics Letters*, vol. 36, pp. 190–192, Feb. 1980.

[67] P. Ressel, G. Erbert, U. Zeimer, K. Hausler, G. Beister, B. Sumpf, A. Klehr, and G. Tränkle, "Novel passivation process for the mirror facets of Al-free active-region high-power semiconductor diode lasers," *IEEE Photonics Technology Letters*, vol. 17, pp. 962–964, May 2005.

[68] "ISO 11146-1:2005(en) Lasers and laser-related equipment — Test methods for laser beam widths, divergence angles and beam propagation ratios - Part 1." https://www.iso.org/obp/ui/#iso:std:iso:11146:-1:ed-1:v1:en/, Accessed 15 July 2017.

[69] M. Maeda, T. Hirata, M. Suehiro, M. Hihara, A. Yamaguchi, and H. Hosomatsu, "Photonic Integrated Circuit Combining Two GaAs Distributed Bragg Reflector Laser Diodes for Generation of the Beat Signal," *Japanese Journal of Applied Physics*, vol. 31, p. L183, Feb. 1992.

[70] R. K. Price, V. B. Verma, K. E. Tobin, V. C. Elarde, and J. J. Coleman, "Y-Branch Surface-Etched Distributed Bragg Reflector Lasers at 850 nm for Optical Heterodyning," *IEEE Photonics Technology Letters*, vol. 19, pp. 1610–1612, Oct. 2007.

[71] Masahiro Uemukai, H. Ishida, A. Ito, T. Suhara, H. Kitajima, A. Watanabe, and H. Kan, "Integrated AlGaAs Quantum-Well Ridge-Structure Two-Wavelength Distributed Bragg Reflector Laser for Terahertz Wave Generation," *Japanese Journal of Applied Physics*, vol. 51, p. 020205, Feb. 2012.

[72] A. P. Shreve, N. J. Cherepy, and R. A. Mathies, "Effective Rejection of Fluorescence Interference in Raman Spectroscopy Using a Shifted Excitation Difference Technique," *Applied Spectroscopy*, vol. 46, pp. 707–711, Apr. 1992.

[73] M. Maiwald, B. Eppich, J. Fricke, A. Ginolas, F. Bugge, B. Sumpf, G. Erbert, and G. Tränkle, "Dual-Wavelength Y-Branch Distributed Bragg Reflector Diode Laser at 785 Nanometers for Shifted Excitation Raman Difference Spectroscopy," *Applied Spectroscopy*, vol. 68, pp. 838–843, Aug. 2014.

[74] M. Maiwald, J. Fricke, A. Ginolas, J. Pohl, B. Sumpf, G. Erbert, and G. Tränkle, "Dual-wavelength monolithic Y-branch distributed Bragg reflection diode laser at 671 nm suitable for shifted excitation Raman difference spectroscopy," *Laser & Photonics Reviews*, vol. 7, pp. L30–L33, July 2013.

[75] B. Sumpf, M. Maiwald, A. Müller, J. Fricke, P. Ressel, F. Bugge, G. Erbert, and G. Tränkle, "Comparison of two concepts for dual-wavelength DBR ridge waveguide diode lasers at 785 nm suitable for shifted excitation Raman difference spectroscopy," *Applied Physics B*, vol. 120, pp. 261–269, Aug. 2015.

[76] P. L. Liu, B. J. Li, P. J. Cressman, J. R. Debesis, and S. Stoller, "Comparison of measured losses of Ti:LiNbO3 channel waveguide bends," *IEEE Photonics Technology Letters*, vol. 3, pp. 755–756, Aug. 1991.

[77] R. B. Swint, T. S. Yeoh, V. C. Elarde, J. J. Coleman, and M. S. Zediker, "Curved waveguides for spatial mode filters in semiconductor lasers," *IEEE Photonics Technology Letters*, vol. 16, pp. 12–14, Jan. 2004.

[78] R. G. Walker, N. I. Cameron, Y. Zhou, and S. J. Clements, "Optimized Gallium Arsenide Modulators for Advanced Modulation Formats," *IEEE Journal of Selected Topics in Quantum Electronics*, vol. 19, pp. 138–149, Nov. 2013.

[79] "FIMMWAVE Photon Design Ltd., Photon Design." http://www.photond.com/, Accessed 24 May 2018.

[80] J. S. Gu, P. A. Besse, and H. Melchior, "Method of lines for the analysis of the propagation characteristics of curved optical rib waveguides," *IEEE Journal of Quantum Electronics*, vol. 27, pp. 531–537, Mar. 1991.

[81] R. C. Alferness, U. Koren, L. L. Buhl, B. I. Miller, M. G. Young, T. L. Koch, G. Raybon, and C. A. Burrus, "Broadly tunable InGaAsP/InP laser based on a vertical coupler filter with 57-nm tuning range," *Applied Physics Letters*, vol. 60, pp. 3209–3211, June 1992.

[82] S. L. Woodward, U. Koren, B. I. Miller, M. G. Young, M. A. Newkirk, and C. A. Burrus, "A DBR laser tunable by resistive heating," *IEEE Photonics Technology Letters*, vol. 4, pp. 1330–1332, Dec. 1992.

[83] B. Sumpf, J. Kabitzke, J. Fricke, P. Ressel, A. Müller, M. Maiwald, and G. Tränkle, "785nm dual-wavelength Y-branch DBR-RW diode laser with electrically adjustable wavelength distance between 0 nm and 2 nm," in *Novel In-Plane Semiconductor Lasers XVI*, vol. 10123, p. 101230T, International Society for Optics and Photonics, Feb. 2017.

[84] "PH Photo Detectors and Other Products | Gentec-EO." https://www.gentec-eo.com/products/photo-detectors/PH/, Accessed 16 June 2018.

[85] T. F. Krauss, G. Hondromitros, B. Vögele, and R. M. D. L. Rue, "Broad spectral bandwidth semiconductor lasers," *Electronics Letters*, vol. 33, pp. 1142–1143, June 1997.

[86] V. K. Kononenko, I. S. Manak, and S. V. Nalivko, "Design and characteristics of widely tunable quantum-well laser diodes," *Spectrochimica Acta Part A: Molecular and Biomolecular Spectroscopy*, vol. 55, pp. 2091–2096, Sept. 1999.

[87] M. J. Hamp and D. T. Cassidy, "Critical design parameters for engineering broadly tunable asymmetric multiple-quantum-well lasers," *IEEE Journal of Quantum Electronics*, vol. 36, pp. 978–983, Aug. 2000.

[88] N. Dutta and P. Deimel, "Optical properties of a GaAlAs superluminescent diode," *IEEE Journal of Quantum Electronics*, vol. 19, pp. 496–498, Apr. 1983.

[89] P.-J. Rigole, S. Nilsson, L. Backbom, T. Klinga, J. Wallin, B. Stalnacke, E. Berglind, and B. Stoltz, "114-nm wavelength tuning range of a vertical grating assisted codirectional coupler laser with a super structure grating distributed Bragg reflector," *IEEE Photonics Technology Letters*, vol. 7, pp. 697–699, July 1995.

[90] P. Della Casa, O. Brox, A. Knigge, B. Sumpf, M. Tawfieq, H. Wenzel, M. Weyers, and G. Tränkle, "Integration of active, passive and buried-grating sections for a GaAs-based, widely tunable laser with sampled grating Bragg reflectors," in *Compound Semiconductor Week 2017 (CSW)*, (Berlin, Germany), May 2017.

[91] V. Jayaraman, Z. M. Chuang, and L. A. Coldren, "Theory, design, and performance of extended tuning range semiconductor lasers with sampled gratings," *IEEE Journal of Quantum Electronics*, vol. 29, pp. 1824–1834, June 1993.

[92] "According to internal results at the Ferdinand-Braun-Institute, 05 September 2017."

[93] U. Bandelow and U. Leonhardt, "Light propagation in one-dimensional lossless dielectrica: Transfer matrix method and coupled mode theory," *Optics Communications*, vol. 101, pp. 92–99, Aug. 1993.

[94] A. Yariv, "Coupled-mode theory for guided-wave optics," *IEEE Journal of Quantum Electronics*, vol. 9, pp. 919–933, Sept. 1973.

[95] H. Wenzel, "Green's function based simulation of the optical spectrum of multisection lasers," *IEEE Journal of Selected Topics in Quantum Electronics*, vol. 9, pp. 865–871, May 2003.

[96] Ferdinand-Braun-Institut, "Design Tools - Optoelectronics Department | Ferdinand-Braun-Institut." https://www.fbh-berlin.com/research/photonics/optoelectronics-department/design-tools, Accessed 30 August 2018.

[97] D. J. Klotzkin, *Introduction to Semiconductor Lasers for Optical Communications*, vol. Chapter 9.4. New York, NY: Springer New York, 2014.

[98] A. Klehr, H. Wenzel, J. Fricke, F. Bugge, A. Liero, T. Hoffmann, G. Erbert, and G. Tränkle, "Generation of spectrally-stable continuous-wave emission and ns pulses at 800 nm and 975 nm with a peak power of 4 W using a distributed Bragg reflector laser and a ridge-waveguide power amplifier," in *Novel In-Plane Semiconductor Lasers XIV*, vol. 9382, p. 93821I, International Society for Optics and Photonics, Mar. 2015.

[99] P. Crump, O. Brox, F. Bugge, J. Fricke, C. Schultz, M. Spreemann, B. Sumpf, H. Wenzel, and G. Erbert, "Chapter 2 - High-Power, High-Efficiency Monolithic Edge-Emitting GaAs-Based Lasers with Narrow Spectral Widths," in *Semiconductors and Semimetals* (J. J. Coleman, A. C. Bryce, and C. Jagadish, eds.), vol. 86 of *Advances in Semiconductor Lasers*, pp. 49–91, Elsevier, Jan. 2012.

[100] B. Sumpf, K. H. Hasler, P. Adamiec, F. Bugge, F. Dittmar, J. Fricke, H. Wenzel, M. Zorn, G. Erbert, and G. Tränkle, "High-Brightness Quantum Well Tapered Lasers," *IEEE Journal of Selected Topics in Quantum Electronics*, vol. 15, pp. 1009–1020, May 2009.

[101] A. Müller, J. Fricke, F. Bugge, O. Brox, G. Erbert, and B. Sumpf, "DBR tapered diode laser with 12.7 W output power and nearly diffraction-limited, narrowband emission at 1030 nm," *Applied Physics B*, vol. 122, p. 87, Apr. 2016.

[102] H. Odriozola, J. M. G. Tijero, L. Borruel, I. Esquivias, H. Wenzel, F. Dittmar, K. Paschke, B. Sumpf, and G. Erbert, "Beam Properties of 980-nm Tapered Lasers With Separate Contacts: Experiments and Simulations," *IEEE Journal of Quantum Electronics*, vol. 45, pp. 42–50, Jan. 2009.

[103] "BeamXpertDESIGNER." https://www.beamxpert.de/, Accessed 03 July 2018.

[104] F. L. Pedrotti, L. M. Pedrotti, and L. S. Pedrotti, "Introduction to Optics, 3rd Edition," in *Introduction to Optics*, pp. 396–418, Pearson, 3 ed.

[105] "SmarAct – Perfect Motion." http://www.smaract.com/, Accessed 05 June 2018, Feb. 2016.

[106] M. Braune, M. Maiwald, B. Eppich, O. Brox, A. Ginolas, B. Sumpf, and G. Tränkle, "Design and Realization of a Miniaturized DFB Diode Laser-Based SHG Light Source With a 2-nm Tunable Emission at 488 nm," *IEEE Transactions on Components, Packaging and Manufacturing Technology*, vol. 7, pp. 720–725, May 2017.

[107] "Block engineering - Quantum Cascade Lasers for OEM Customers." http://www.blockeng.com/products/miniqcl.html/, Accessed 19 July 2018.

[108] "Clear Optical Path USAF Target | Edmund Optics." https://www.edmundoptics.com/f/clear-optical-path-usaf-target/13107/, Accessed 19 July 2018.

[109] L. Huot, P. M. Moselund, P. Tidemand-Lichtenberg, and C. Pedersen, "Electronically delay-tuned upconversion cross-correlator for characterization of mid-infrared pulses," *Optics Letters*, vol. 43, p. 2881, June 2018.

[110] W. Wei, H. Deng, and J. He, "GaAs/AlGaAs-Based 870-nm-Band Widely Tunable Edge-Emitting V-Cavity Laser," *IEEE Photonics Journal*, vol. 5, pp. 1501607–1501607, Oct. 2013.

[111] V. Jayaraman, G. D. Cole, M. Robertson, C. Burgner, D. John, A. Uddin, and A. Cable, "Rapidly swept, ultra-widely-tunable 1060 nm MEMS-VCSELs," *Electronics Letters*, vol. 48, pp. 1331–1333, Oct. 2012.

[112] M. C. Y. Huang, Y. Zhou, and C. J. Chang-Hasnain, "A nanoelectromechanical tunable laser," *Nature Photonics*, vol. 2, pp. 180–184, Mar. 2008.

[113] S. R. Chinn, E. A. Swanson, and J. G. Fujimoto, "Optical coherence tomography using a frequency-tunable optical source," *Optics Letters*, vol. 22, pp. 340–342, Mar. 1997.

[114] "CTL 950 - 950 Continuously Tunable - TOPTICA Photonics." https://www.toptica.com/products/tunable-diode-lasers/ecdl-dfb-lasers/ctl/, Accessed 16 Aug. 2018, July 2017.

[115] "850 - Continuously Tunable Laser - TOPTICA Photonics." https://www.toptica.com/products/tunable-diode-lasers/ecdl-dfb-lasers/dl-pro/, Accessed 16 Aug. 2018.

[116] "TEC-420-0976-2000 - Amplified Tunable Diode Laser - Sacher Lasertechnik." https://www.sacher-laser.com/home/scientific-lasers/tapered_amplifiers/amplified_tunable_laser/mopa-amplified-tunable-diode-laser_system_servalplus.html, Accessed 16 Aug. 2018.

[117] A. F. Nieuwenhuis, C. J. Lee, B. Sumpf, P. J. M. van der Slot, G. Erbert, and K.-J. Boller, "One-Watt level mid-IR output, singly resonant, continuous-wave optical parametric oscillator pumped by a monolithic diode laser," *Optics Express*, vol. 18, pp. 11123–11131, May 2010.

[118] "Surface Profilometry and Surface Metrology - Stylus Profilometer and Surface Measurement | KLA Tencor." https://www.kla-tencor.com/surface-profilometry-and-metrology.html, Accessed 27 Aug. 2018.

Innovationen mit Mikrowellen und Licht
Forschungsberichte aus dem Ferdinand-Braun-Institut, Leibniz-Institut für Höchstfrequenztechnik

Herausgeber: Prof. Dr. G. Tränkle, Prof. Dr.-Ing. W. Heinrich

Cuvillier Verlag
Internationaler wissenschaftlicher Fachverlag

Innovationen mit Mikrowellen und Licht
Forschungsberichte aus dem Ferdinand-Braun-Institut, Leibniz-Institut für Höchstfrequenztechnik

Herausgeber: Prof. Dr. G. Tränkle, Prof. Dr.-Ing. W. Heinrich

Cuvillier Verlag
Internationaler wissenschaftlicher Fachverlag

Innovationen mit Mikrowellen und Licht
Forschungsberichte aus dem Ferdinand-Braun-Institut, Leibniz-Institut für Höchstfrequenztechnik

Herausgeber: Prof. Dr. G. Tränkle, Prof. Dr.-Ing. W. Heinrich

Cuvillier Verlag
Internationaler wissenschaftlicher Fachverlag

Innovationen mit Mikrowellen und Licht
Forschungsberichte aus dem Ferdinand-Braun-Institut, Leibniz-Institut für Höchstfrequenztechnik

Herausgeber: Prof. Dr. G. Tränkle, Prof. Dr.-Ing. W. Heinrich

Band 30:
Christian Fiebig
Diodenlaser mit Trapezstruktur und hoher Brillanz für die Realisierung einer Frequenzkonversion auf einer mikro-optischen Bank
ISBN: 978-3-95404-690-4, 26,30 EUR, 140 Seiten

Band 31:
Viola Küller
Versetzungsreduzierte AlN- und AlGaN-Schichten als Basis für UV LEDs
ISBN: 978-3-95404-741-3, 34,40 EUR, 164 Seiten

Band 32:
Daniel Jedrzejczyk
Efficient frequency doubling of near-infrared diode lasers using quasi phase-matched waveguides
ISBN: 978-3-95404-958-5, 27,90 EUR, 134 Seiten

Band 33:
Sylvia Hagedorn
Hybrid-Gasphasenepitaxie zur Herstellung von Aluminiumgalliumnitrid
ISBN: 978-3-95404-985-1, 38,00 EUR, 176 Seiten

Band 34:
Alexander Kravets
Advanced Silicon MMICs for mm-Wave Automotive Radar Front-Ends
ISBN: 978-3-95404-986-8, 31,90 EUR, 156 Seiten

Band 35:
David Feise
Longitudinale Modenfilter für Kantenemitter im roten Spektralbereich
ISBN: 978-3-7369-9116-3, 39,20 EUR, 168 Seiten

Band 36:
Ksenia Nosaeva
Indium phosphide HBT in thermally optimized periphery for applications up to 300GHZ
ISBN: 978-3-7369-287-0, 42,00 EUR, 154 Seiten

Band 37:
Muhammad Maruf Hossain
Signal Generation for Millimeter Wave and THZ Applications in InP-DHBT and InP-on-BiCMOS Technologies
ISBN: 978-3-7369-9335-8, 35,60 EUR, 136 Seiten

Band 38:
Sirinpa Monayakul
Development of Sub-mm Wave Flip-Chip Interconnect
ISBN: 978-3-7369-9410-2, 44,00 EUR, 146 Seiten

Band 39:
Moritz Brendel
Charakterisierung und Optimierung von (Al, Ga) N-basierten UV-Photodetektoren
ISBN: 978-3-7369-9465-2, 49,90 EUR, 196 Seiten

Band 40:
Erdenetsetseg Luvsandamdin
Development of micro-integrated diode lasers for precision quantum optics experiments in space
ISBN: 978-3-7369-9479-9, 39,00 EUR, 126 Seiten

Cuvillier Verlag
Internationaler wissenschaftlicher Fachverlag

Innovationen mit Mikrowellen und Licht
Forschungsberichte aus dem Ferdinand-Braun-Institut, Leibniz-Institut für Höchstfrequenztechnik

Herausgeber: Prof. Dr. G. Tränkle, Prof. Dr.-Ing. W. Heinrich

Cuvillier Verlag
Internationaler wissenschaftlicher Fachverlag

www.ingramcontent.com/pod-product-compliance
Lightning Source LLC
Chambersburg PA
CBHW060449240326
41598CB00088B/4308